Graphing Calculator Guide for the TI-83

FUNCTIONS
MODELING
CHANGE

A PREPARATION FOR CALCULUS

Graphing Calculator Guide for the TI-83

Carl Swenson
Seattle University

FUNCTIONS MODELING CHANGE

A PREPARATION FOR CALCULUS

Second Edition

Eric Connally
Harvard University Extension

Deborah Hughes-Hallett
University of Arizona

Andrew M. Gleason
Harvard University

et al.

WILEY

John Wiley & Sons, Inc.

COVER PHOTO: ©Nick Wood

 This material is based upon work supported by the National Science Foundation under Grant No. DUE-9352905. Opinions expressed are those of the authors and not necessarily those of the Foundation.

To order books or for customer service call 1-800-CALL-WILEY (225-5945).

ISBN 0-471-44789-7

Printed in the United States of America

10 9 8 7 6 5 4 3 2 1

Printed and bound by Bradford & Bigelow, Inc.

PREFACE

The purpose of this book is to apply the features of the TI-83 graphing calculators to understand precalculus. The examples are taken from the precalculus book, *Functions Modeling Change, 2nd Edition* by Connally, Hughes-Hallett, Gleason, et al. I would like to acknowledge and thank the Calculus Consortium for Higher Education (CCHE) and John Wiley & Sons, Inc. for permission to freely use the text examples.

To the student

Using a graphing calculator can be both fun and frustrating. A healthy attitude when you get frustrated is to step back and say, "Isn't that interesting that it doesn't work." Figuring out how things work can be fun. If you get too frustrated, then it is time to ask a friend or the instructor for help.

Chapter Zero (Getting Started,) gives you a basic orientation to make you comfortable with how your calculator works. The remaining chapters, One to Twelve, correspond to the Table of Contents of the text *Functions Modeling Change*.

If you have a specific calculator question, your *TI–Guidebook* provides a resource that explains each feature briefly, and usually gives an example. (If you do not have the *TI–Guidebook*, it can be downloaded from the TI website.) Remember that the *TI-Guidebook* is like a dictionary: there is no story line or context. My intent here is to introduce the calculator features that you need for precalculus, so features are explained in the context of a precalculus example. Some calculator features that are less important to precalculus are not mentioned at all. The mathematical content drives this presentation, not the calculator features.

I have included tips about such things as shortcuts, warnings, and related ideas. I hope you find them useful.

Tip: Don't use technology in place of thinking.

To the instructor

It has been my aim to write this manual so that you can assign students to read it and practice the examples independently. It can also be used for in-class presentations. This manual focuses on precalculus examples, not on the calculator features. It has been my experience that the problems of teaching with this technology are greatly reduced by requiring a single standard calculator for a course. A mix of TI-83 and TI-82 calculators can be tolerable, but even this can cause occasional turbulence.

Will these materials meet the needs of all your students? Of course not. There will still be the zealous students who want to write extensive programs and the anxious students who want all the buttons pressed for them. These materials are aimed at the middle, giving enough guidance so that most students will be able to work through an example without assistance, but not so specific as to create a mindless exercise in pressing keys in the right order.

Tip: I strongly recommend using an overhead-projected calculator as a tool in your classroom. In most cases the TI Volume Purchase Plan can provide you with this package for free. I also recommend the computer-to-calculator cable, *TI-Graph Link* ™, and the software *TI-Connect*.

Thanks

I would like to thank Seattle University for its continued support of office and computer facilities. And special thanks to Troy Heffner for his willingness and ability to answer software questions. I would like to acknowledge the debt I owe to Ann Davidian for her exuberent eagle-eyed editing of the first edition. Thanks to the folks at Wiley who have been a great help in answering questions and making arrangements.

— Carl Swenson, *swenson@seattleu.edu*

TABLE OF CONTENTS

CHAPTER ZERO

BEFORE STARTING

Basic Calculations

Calculators have developed from the fingers, to the abacus, to the slide rule, to the scientific calculator, and now we have the graphing calculator. In this chapter you will get a gentle introduction to using the TI-83 graphing calculator and you will see how to make some simple numerical calculations. If you have used a graphing calculator before, you may only need to skim this chapter. The *TI–Guidebook* should also be consulted if you are having difficulty getting started. In this book you will find the references to TI calculator keys and menu choices written in the TI-83 font. The TI-83 font looks like this.

A Note for TI-82 Users: If you are a TI-82 user, the screens will, as a rule, be the same as those in this book. The TI-82 has fewer features than the TI-83, so some of its menus are shorter. With few exceptions, this manual can be used for the TI-82.

A Note for TI-83 Users: There are currently three models of the TI-83: non-Plus, Plus, and Plus Silver Edition. In this manual it makes little difference which model is used. The main difference between models is the calculation speed and the storage capacity. The non-Plus model cannot download and run special application programs called apps.

Getting Started with Essential Keys

The ON key
Study the keyboard and press the ON key in the lower left-hand corner. You should see a blinking rectangular cursor. If not, then you may need to set the screen contrast. Even if the cursor is showing, it is a good idea to know how to adjust the screen contrast.

The 2nd key adjusts the screen contrast
As you use the calculator, the battery wears down and it becomes necessary to adjust the screen. Also, you may need to adjust the screen contrast for different lighting environments. Press and release the yellow key marked 2nd and then the blue up-arrow key in the upper right of the keypad. By repeating this sequence, the screen darkens. The screen can be lightened by repeating this sequence but by using the down-arrow instead of the up-arrow key. A momentary value (between 0 and 9) flashes in the upper right corner of the screen telling you the battery status (9 is close to replacement time). If the setting is too low, the cursor does not show; and if it is too high, the screen is dark as night.

If you take a break and come back later, the cursor disappears for a different reason. The calculator goes to sleep and turns itself off after a few minutes of no activity. Just press the ON key and it wakes up at the same place it turned off: no memory loss. The 2ndOFF key turns off the calculator immediately. (The notation, 2ndOFF, means press the 2nd key and then press the key that has OFF written above it.)

Tip: Sometimes broken calculators can be fixed by reinserting the batteries correctly.

Arithmetic Calculations on the Home Screen

Use technology on known results before trying complex examples. Our first use of the calculator will be to do arithmetic on the home screen. Many different screens are shown later, but this *home* screen is where you do calculations. A graphing calculator has a distinct calculation advantage over a scientific calculator because it shows multiple lines and has entry recall.

Type some calculation for which you know the answer, say 8*9, and press the **ENTER** key; the result appears on the right side of the screen. You can see successive entries of three known multiplications. Had they not been seen all together, you may not have noticed an interesting pattern: the sum of the answer digits is always 9 (i.e., $7 + 2 = 9$, $6 + 3 = 9$, and $5 + 4 = 9$). On a graphing calculator, your results flow down the screen as you work.

```
8*9
              72
7*9
              63
6*9
              54
■
```

The white numeric keys and the blue operation keys are used for simple arithmetic calculations. Type

3 ÷ 2 ENTER

Be aware that the symbol on the divide key, ÷, is different from the divide symbol (/) that appears on the screen. Take special note that the negation key ⟨-⟩ must be distinguished from the blue ─ subtraction key. One of the most common errors is interchanging the use of the subtraction key ─ and the negation key ⟨-⟩. Give it a moment s thought: subtraction requires two numbers, while negation works on a single number. Press the following four keys:

3 ─ 2 ENTER

and then

3 ⟨-⟩ 2 ENTER

```
3/2
             1.5
3-2
               1
3-2■
```

Look carefully at the previous screen that shows the two similar symbols: subtraction, which is longer and centered, and negation, which is shorter and raised. This second entry gives you a syntax error shown here. Error screens replace the work screen and tell you briefly what is wrong, and you are forced to respond with **1:Quit** or **2:Goto**. The **Goto** option is usually the best choice because it will go to the error location and allow you to correct it.

```
ERR:SYNTAX
1■Quit
2:Goto
```

Tip: Sequences of calculator screens start and stop with a heavy top and bottom border (as shown above). These borders are meant to mark the beginning and the end of a sequence; this can be especially helpful when a sequence does not appear on a single page.

Scientific Keys

Next we test the scientific keys. These give values for many expressions used in science, such as the common log (**LOG**) and natural log (**LN**) function. The **TAN**, **COS**, and **SIN** keys are the standard trigonometric functions. These functions appear on the screen in lower case followed by a left parenthesis. First, however, we examine the taking of powers.

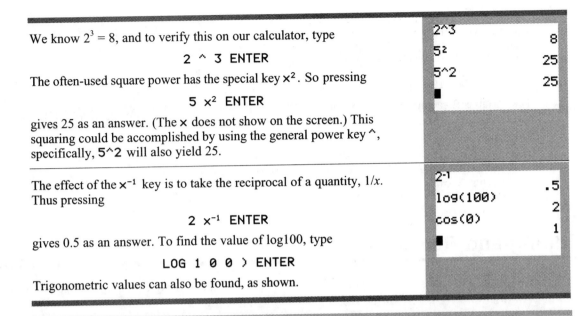

We know $2^3 = 8$, and to verify this on our calculator, type

2 ^ 3 ENTER

The often-used square power has the special key x^2. So pressing

5 x^2 ENTER

gives 25 as an answer. (The x does not show on the screen.) This squaring could be accomplished by using the general power key $^$, specifically, 5^2 will also yield 25.

The effect of the x^{-1} key is to take the reciprocal of a quantity, $1/x$. Thus pressing

2 x^{-1} ENTER

gives 0.5 as an answer. To find the value of log100, type

LOG 1 0 0) ENTER

Trigonometric values can also be found, as shown.

Tip: The right parenthesis in an expression like log(100) is not actually required, but it is a bad habit to leave it off.

Magic Tricks to Change the Keyboard

How can we do more with the basic keys? The trick is multiple-state keys. The indicator of the keyboard state is shown by the cursor. The standard is a solid black square. The first change state key is the yellow **2nd** key. After pressing **2nd** once, the inside of the cursor will show an up arrow Presto! All the keys now have a new meaning. These   meanings are indicated in yellow, just above and to the left of each key. We have already used the **2nd** key to adjust the screen contrast and to turn the calculator off.

Practicing the 2nd key on the greatest equation ever written

Five symbols, 0, 1, e, „, and i, are frequently used in mathematics. Incredible as it might seem, they can be related by a single equation:

$$e^{\pi \cdot i} + 1 = 0.$$

You can practice using the 2nd key for e, „, and i by entering

2nde^x 2ndπ * 2nd$i$) + 1

(The TI-82 has no i key and can t do this equation.)

Using ALPHA to store values

The second change state key is the green **ALPHA** key; as its name indicates, it is used to enter alphabetic letters such as variables, and it has a secondary role as the **Solver** activate key. (This feature will be discussed in Chapter 2.)

The store key, **STO▸**, which appears on the screen as →, is used to store a numeric value into a letter variable. If you want to repeatedly use the value of (log(100)+1)/2 in calculations, then you enter:

    (**LOG** 100) + 1) / 2 **STO▸ ALPHA** A **ENTER**

The variable A can now be used in computations, as shown.

```
(log(100)+1)/2→A
                 1.5
3*A
                 4.5
A+A
                   3
■
```

Tip: The cursor box changes if the **ALPHA** or **2nd** keys are in effect. The **ALPHA** and **2nd** keys work as toggles: if you press one by mistake and turn on a state, just press the key again to turn it off.

Editing and Recall

We all make mistakes; correcting them on a graphing calculator is relatively easy. Use the arrow keys to navigate the screen and type over your errors.

Corrections in Longer Expressions

The single most popular error (can errors be popular?) among new users is the failure to use parentheses when needed. This is a serious error because the calculator does not stop and alert you with an error screen; instead, it gives you the correct answer to a question you are not asking. There is a prescribed order of operations on your calculator; you can look this up in your *TI–Guidebook* for details if you have questions about this order.

Suppose you want to add 2 and 6 and then divide by 8. We don t need a calculator to tell us the expression has value 1. But if you enter 2 + 6 / 8, the answer is 2.75. You can figure out that the calculator divided 6 by 8 first and then added that to 2. This was not what we wanted. We need to use parentheses to ensure that we are evaluating the correct expression. Now try (2 + 6) / 8 and get an answer of 1 as expected.

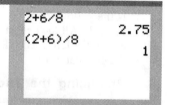

```
2+6/8
              2.75
(2+6)/8
                 1
```

Tip: When you get an unexpected result, go back and check parentheses. Be generous; adding extra parentheses doesn t hurt.

Correction keys: DEL, CLEAR, and 2ndINS

While on the home screen, you can use the blue arrow keys to move forward and backward. (The up arrow takes you to the beginning of the line, the down arrow to the end.) Press the **DEL** key to delete the character in the cursor box. Use **CLEAR** to delete the whole entry line; if the entry line is already clear, then the whole screen clears. Thus, pressing **CLEAR** twice always clears the screen. If you need to insert one or more characters, you can move (using the arrow keys) to the location at which you want to insert and press the **2nd INS** key. The cursor appears as an underline when the insert mode is in effect. Pressing any arrow key turns off the insert mode.

Tip: Unfortunately, there is no backspace key such as is found on computer keyboards.

Recalling a Previous Entry Line

Often we see an error after we have pressed ENTER and left the entry line. The 2ⁿᵈENTRY key (above ENTER) places the previous entry line on the screen so that it can be edited for the next calculation. In the last screen example, we had the expression 2+6/8 but then realized we needed to add parentheses. After seeing our mistake on the first line, we could have used 2ⁿᵈENTRY to paste a new copy of the expression on the screen. Then, by using the navigation arrows, we could have inserted the needed parentheses. Note that we would need to use 2ⁿᵈINS since we were not writing over what was there but inserting new parentheses characters.

Deep recall

By repeatedly pressing the 2ⁿᵈENTRY key, many of the previous entry lines are accessible. This is called deep recall and is limited to about 128 characters. Using a previous entry overwrites the current entry; you do not go to a new line. A previous entry cannot be inserted into the command line without erasing what was already there   it is all or nothing.

Recycling a Previous Answer

There is another keystroke-saving feature that is quite handy: the 2ⁿᵈANS key. The 2ⁿᵈANS places the variable Ans on the screen and uses the previous answer as its value when calculated.

To find the difference in area between a 10- and 12-inch pizza, we first find the 10-inch area. Next we use 2ⁿᵈENTRY to place the 10-inch calculation back on the screen. We arrow left and change the 10 to 12.	π(10)^2 314.1592654 π(1■)^2

Arrow back to the right, add the subtraction sign, press 2ⁿᵈANS, and then press ENTER to find the answer.	π(10)^2
	          314.1592654
	π(12)^2-Ans
	          138.2300768

If you start an expression by pressing an operation key, +, −, *, /, ^, the calculator assumes the first number in the calculation is the previous answer, so it puts Ans on the screen without your even pressing the 2ⁿᵈANS key. Press CLEAR and this sequence of keystrokes:
 	
    2 ENTER + 3 ENTER ENTER ENTER ENTER ENTER
	2
	                    2
	Ans+3
	                    5
	                    8
	                   11
	                   14

Tip: Work smart. Use 2ⁿᵈENTRY and 2ⁿᵈANS.

Menus

We have already seen one example of a menu when we committed the error of using the negation sign in place of the subtraction symbol. There were only two choices, but the principle is the same in other cases. Menus are lists of choices that would be far too numerous to have available on the keypad.

Press **MATH** and you will see a menu screen that takes the place of the home screen. Notice that the options are numbered and you can choose by number or by arrowing to a choice and pressing **ENTER**. All the selections may mean little to you at this point at this point.

```
MATH  NUM CPX PRB
1▪▶Frac
2:▶Dec
3:3
4:3√(
5:×√
6:fMin(
7↓fMax(
```

The top row shows four menus: **MATH NUM CPX PRB**. Press the right-arrow key to get the **NUM** menu. The **abs(** is our choice, so we press **ENTER** (or **1**).

```
MATH NUM CPX PRB
1▪abs(
2:round(
3:iPart(
4:fPart(
5:int(
6:min(
7↓max(
```

Now the home screen reappears and **abs(** has been pasted onto the home screen. If the home screen is not cleared, the pasting takes place at the location of the cursor, even in the middle of any expression.

```
abs(█


```

In case you hadn t guessed, **abs** is the absolute value function, which makes numbers positive. Here we see it work. Note that we used the negation sign, not the subtraction key. (The subtraction key will cause an error.)

```
abs(-6)
              6
```

Tip:　Most accomplished users use the number choice as the fastest method of choice in menus. However, if the choice is the first item, then **ENTER** is best.

Tip:　A menu may not show all its items. An arrow in place of the colon indicates that unseen items are available in the arrow s direction.

Tip:　Menus are wrap-around so you can press the up arrow once to get to the last item in a list. A choice of menus on the top line wraps from left or right by using the left or right arrows.

A Catalog of Items

Not knowing in which menu an item is hiding was frustrating on the TI-82, so an alphabetical list feature has been added to the TI-83.

The **2ⁿᵈ CATALOG** key shows an alphabetic list of items. The list is long, but the **a** in the upper right corner is a reminder that you are in **ALPHA** mode and you can conveniently jump closer to the desired item by pressing the key of its first letter.

Changing the Format: MODE

You can control the output format of numerical calculations so that they are all shown in scientific notation. Or, if you are doing a business application, you might want money answers to come out rounded to two decimal places for the dollar-and-cents format.

The **MODE** key allows you to check and change formats. The default settings are all on the left of the screen so a quick glance tells you if any settings have been changed.

```
Normal Sci Eng
Float 0123456789
Radian Degree
Func Par Pol Seq
Connected Dot
Sequential Simul
Real a+bi re^θi
Full Horiz G-T
```

To change a setting, use the down arrow to reach the desired line and then use the right arrow to move across to the desired setting. You must now press **ENTER** to make the change. Press **2ndQUIT** to return to your home screen.

```
Normal Sci Eng
Float 0123456789
Radian Degree
Func Par Pol Seq
Connected Dot
Sequential Simul
Real a+bi re^θi
Full Horiz G-T
```

To see the difference in the two settings, we twice find the total cost of a $1.60 Cafe Latte in Seattle if the sales tax is 8.6%. First we use the default setting, and then we change the format to two decimals. It should be noted that rounding the result in a display format does not change the stored accuracy. To see this, multiply the answer by 1000 and you see that the fuller decimal accuracy is preserved.

```
1.60*1.086
              1.7376
1.60*1.086
                1.74
Ans*1000
             1737.60
```

The **MODE** setting choices are described in the *TI–Guidebook*, pages 1-9 to 1-12. We will mention other settings as we need them, but unless noted otherwise, all our examples will assume that the default settings are in effect.

Tip: If your output values are in an unexpected or undesirable format, check the **MODE** settings. If you are having trouble changing the **MODE** settings, you may have forgotten to press **ENTER** before **2ndQUIT.**

Notes:

CHAPTER ONE

FUNCTIONS, LINES AND CHANGE

1.1 Functions and Function Notation

Definition: A *function* is a rule that takes certain values as inputs and assigns to each input value exactly one output value. The use of functions and functional notation is the primary focus of precalculus since it is vital to success in calculus.

In this chapter, we use the calculator to define and evaluate functions, to make tables of values, and to graph functions. In short, we see how to view a function in words, by table, by graph, and by formula (the Rule of Four).

In the previous chapter we used the lower keypad to do our simple numerical calculations. Now we learn to use the Graph/Table keys on the top row, just under the screen. We start with a simple example of a function and show how it can be represented as a formula, graph, and table with the aid of the calculator.

> *Tip:* Most examples from this point on are taken from *Functions Modeling Change* by Connolly et al. Page references to these examples are given in the left margin, enclosed by square brackets. For example, see the [2] under the heading below.

Cricket Chirping Rate

[2] It is a surprising biological fact that most crickets chirp at a rate that increases as the temperature increases. For the snowy tree cricket, the relationship between temperature and the chirp rate is so reliable that this type of cricket is called the thermometer cricket. We can estimate the temperature (in degrees Fahrenheit) by counting the number of chirps in 15 seconds and adding 40. For instance, if we count 20 chirps in 15 seconds, then a good estimate of the temperature is to sum and find 20 + 40 = 60°F.

Words

This example is already given in words, but we might note that rates are most often given in terms of a single unit, so we can think of the chirp rate in terms of the number of chirps per minute. In this case, since 15 seconds is 1/4 of a minute, we could count the number of chirps per minute and multiply by 1/4.

Formula

From the above discussion we can write a formula for the temperature in terms of the number of chirps:

$$T = \frac{1}{4}R + 40.$$

In this formula we identify R as the independent variable and T as the dependent variable. To enter the formula in the calculator, we translate the independent variable to X and the dependent variable to Y.

> *Tip:*   No matter what symbols are used in a function formula, to be used on the TI they must be transformed to have a Y variable equal to an expression using the variable X.

Graph

The next representation is a graph of the function. Like using a camera this requires making a decision about what part is seen. We set a window of values to give us a good picture. Later we find that the calculator also has some preset window commands that act like an automatic focus.

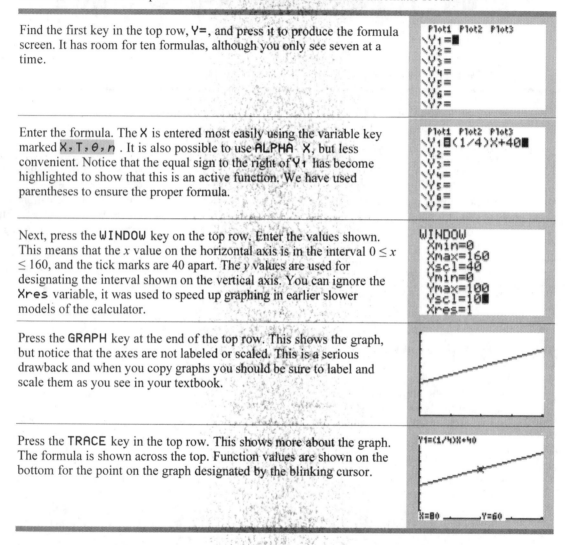

Find the first key in the top row, Y=, and press it to produce the formula screen. It has room for ten formulas, although you only see seven at a time.

Enter the formula. The X is entered most easily using the variable key marked X,T,θ,n . It is also possible to use ALPHA X, but less convenient. Notice that the equal sign to the right of Y₁ has become highlighted to show that this is an active function. We have used parentheses to ensure the proper formula.

Next, press the WINDOW key on the top row. Enter the values shown. This means that the x value on the horizontal axis is in the interval $0 \le x \le 160$, and the tick marks are 40 apart. The y values are used for designating the interval shown on the vertical axis. You can ignore the Xres variable, it was used to speed up graphing in earlier slower models of the calculator.

Press the GRAPH key at the end of the top row. This shows the graph, but notice that the axes are not labeled or scaled. This is a serious drawback and when you copy graphs you should be sure to label and scale them as you see in your textbook.

Press the TRACE key in the top row. This shows more about the graph. The formula is shown across the top. Function values are shown on the bottom for the point on the graph designated by the blinking cursor.

Table

The last representation of the function creates a table of values. Like the graph, we set the starting x value and the step size. There are some more advanced features, but they are introduced later.

To make a table of function values the function formula must be first entered as was just done for the graph. Press the 2ⁿᵈTBLSET key in the top row. This allows you to set the initial value of x (TblStart=) and the step size (ΔTbl=). For now we ignore the other settings.

To see the table of function values, press the 2ⁿᵈ TABLE key in the top row. The table can be scrolled up or down to see additional values. To see values for $x = 0, 1, 2, 3, \ldots$ reset the ΔTbl=1.

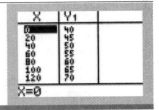

Tip: Recall that to get back to the home screen from a table, press 2ⁿᵈ QUIT.

Tip: Before moving to use advanced features, make sure you know the basic method of entering a formula, graphing a function, and producing a table.

1.2 Rate of Change: Lists and Function Values

The average rate of change tells us about how output values change for a change in input values. For a function $Q = f(t)$ we define

$$\text{Average rate of change} = \frac{\text{Change in } Q}{\text{Change in } t} = \frac{\Delta Q}{\Delta t}$$

Data Lists and Rate of Change

CD and LP Sales

[10] The annual sales (in millions) of compact discs (CDs) and vinyl long playing records (LPs) for selected years between 1982 and 1994 are shown in the table below.

Year	1982	1984	1986	1988	1990	1992	1994
CD sales	0	5.8	53	150	287	408	662
LP sales	244	205	125	72	12	2.3	1.9

What is the rate of change of the annual sales of CDs and LPs between 1982 and 1994?

For CDs, evaluate the quotient, $\dfrac{662 - 0}{12 - 0}$. Simplify to: 662/12. However, such quick simplifications are not always possible, so we find the rate of change over other intervals using a table. Before starting, we note that this is different from earlier examples in that there is no formula to enter in a Y variable. We use the list feature of the calculator to handle this situation where data is given but no function formula is known. There are two different means of entering a list; both are shown.

Press STAT to see the menu shown. Some screens like this one have related menus that can be seen by using the left/right arrows. We use the CALC menu in Section 1.6.

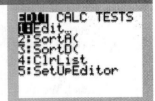

This screen is like a spreadsheet on which to enter numerical lists. There are six numbered lists, L1, L2,..., L6. Only three at a time show. Data can be entered and edited both quickly and easily by using this screen. Remember, just press STAT ENTER. If previous data are in a list, arrow over to use an empty list, or arrow up to the list name and press CLEAR ENTER.

For our CD and LP data we set $t = 0$ for 1982 and enter the t values in L1. Put the CD sales in L2 and the LP sales in L3. Notice that each entry has a name. The one highlighted is L3⟨7⟩, meaning the 7th entry in L3.

Use 2ndQUIT to return to the home screen.

With the lists entered it is easy to set up a difference quotient on the home screen to calculate various rates of change. The list names L1 to L6 are easy to enter (and remember) since they are the 2nd key of 1 to 6. Other rates can be found by using 2ndENTRY and making editing changes. In this example, after the calculation for CDs, the LP values were found by just changing L2 to L3 in two places.

Now that you know about list names and structure, an alternative way of defining a list is on the home screen. Use set brackets { } and commas to create a list and then use the store key, STO▸, followed by one of the six list names.

Tip: Do not press ALPHA and L and 4 in hopes of getting the list name L4. The list name is only accessible through the 2ndL4 key, and the number shows as a subscript on the screen.

Rate of Change from Functions

In a similar fashion you can find the rate of change for a function that is defined by a formula. For a function over the interval $a \le t \le b$, we define

$$\text{Average rate of change} = \frac{\text{Change in } Q}{\text{Change in } t} = \frac{\Delta Q}{\Delta t} = \frac{f(b) - f(a)}{b - a}.$$

[14]   We will find the rate of change of the function $f(x) = x^2$ between $x = 1$ and $x = 3$ and between $x = -2$ and $x = 1$.

Function formulas and rate of change

We could easily enter the function formula in the Y= menu, but we use an alternative (and longer) method. On the home screen you can enclose a formula in quotes and use the STO▸ key to store the formula. (The quote symbol is above the + key.) The hard part is the final pasting of the Y1 symbol. Unlike list names, like L1, there is no keyboard symbol. We must paste Y1 using the following steps.

Press VARS, then right arrow once to show the Y-VARS menu. From this menu press ENTER (or 1) because we want a function name. The next screen shows the function name list: Y₁, Y₂,.... Press 1 to choose Y₁. See the results on the next screen.

```
VARS Y-VARS
1:Function...
2:Parametric...
3:Polar...
4:On/Off...
```

The top line shows the definition line after pasting Y₁ and pressing ENTER. The word Done confirms that the definition was successful. This longer method was shown because the sequence of pasting a Y variable is used over and over. Now the function is evaluated for $x = 3$, by using the common function notation of putting the input in parentheses. This required pasting our function name.

```
"X^2"→Y₁
                  Done
Y₁(3)
                    9
■
```

Once we can evaluate functions, the difference quotient can be entered to find the two rates of change. It is most efficient to use 2ⁿᵈENTRY and be able to use the 2ⁿᵈINS key for inserting an extra character or DEL to remove one. Notice the extra parentheses to ensure correct calculations.

```
(Y₁(3)-Y₁(1))/(3
-1)
                    4
(Y₁(1)-Y₁(-2))/(
1-(-2))
                   -1
```

Tip: Like list names, you cannot press ALPHA and Y and 1 in hopes of pasting the function name Y₁. The function name is only accessible through the sequence of steps: VARS (right-arrow to Y-VARS) ENTER (select).

Tip: Edit long expressions by using 2ⁿᵈENTRY, 2ⁿᵈINS, and DEL, whenever possible.

1.3 Linear Functions

Definition: A *linear* function has a constant rate of change. In this chapter we look at what this means in terms of tables, graphs, and formulas.

Linear Depreciation

[18]   The value of new computer equipment is $20,000, and the value drops at a constant rate so that it is worth $0 after five years. If V is the value of the computer equipment t years after the equipment is purchased, find a formula for V in terms of t.

Formula
The basic formula for a linear equation is

$$\text{Output} = \text{Initial value} + \text{Rate of change} \times \text{Input}.$$

From the above discussion we can find the rate of change:

$$\text{Rate of change} = \frac{\Delta V}{\Delta t} = \frac{-20000}{5} = -4000.$$

The initial value is $20,000, so our linear equation is

$$V = 20000 - 4000t.$$

Graph

The next representation is a graph of a linear function. It should be no surprise that the graph looks like a straight line.

First translate variables and write the formula in Y₁. Then set the viewing **WINDOW** to $0 \le x \le 5$ (Xscl=1) and $0 \le y \le 20000$ (Yscl=4000). **TRACE** shows the equation and graph.

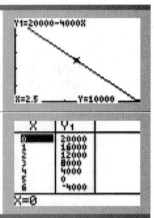

To look at a table, press the **2ⁿᵈTBLSET** and set **TblStart=0** and **△Tbl=1**. Press **2ⁿᵈTABLE** to see the table shown. Notice that for each additional year the value drops by 4000.

Tip: The linear equation is written as $y = mx + b$ or as $y = b + mx$, where m is the constant rate of change and b is the initial value.

We next look at a technique to determine if a function is linear by looking only at its table of values. In the previous case we started with a formula; next we start with a data table.

Checking Tables for Linear Functions

[20] The table below gives values of two functions, p and q. Could either of these functions be linear?

x	50	55	60	65	70
$p(x)$	0.10	0.11	0.12	0.13	0.14
$q(x)$	0.01	0.03	0.06	0.14	0.15

We want to find out if $\dfrac{\Delta p}{\Delta x}$ and $\dfrac{\Delta q}{\Delta x}$ are constant. We put the data into lists, find the differences within a list, and finally see how to divide the differences to give us a list for $\dfrac{\Delta p}{\Delta x}$ and $\dfrac{\Delta q}{\Delta x}$.

Using △List to find differences

We put the three rows of values into the lists **L1, L2, L3**, by using **STAT EDIT ENTER**. If these lists already contain values you can either type over the old values or use **DEL** to delete a single entry. To empty an entire list before entering new data: arrow up to highlight its name, then press **CLEAR** and **ENTER**. (Recall **2ⁿᵈQUIT** returns you to the home screen.)

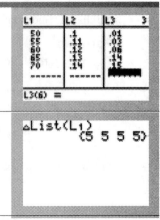

We want the differences *within* a list to be calculated and stored. We use the home screen command:

△List(L1)

The △List is a single command pasted from either the **CATALOG** (under L) or by **LIST OPS 7**.

Use 2ⁿᵈENTRY to save a trip back to the CATALOG and change the list name to L₂. But this time the list extends off the screen. If you want to see it all, you can scroll back and forth with the right/left arrows.

```
∆List(L₁)
          (5 5 5 5)
∆List(L₂)
{.01 .01 .01 .0…
■
```

To answer our question about $\frac{\Delta p}{\Delta x}$ and $\frac{\Delta q}{\Delta x}$ we enter the commands shown to determine that p is linear but q is not. Note that this division is done term-by-term, which is important since all the Δx values from a table may not always be the same as they are in this example.

```
∆List(L₂)/∆List(
L₁)
{.002 .002 .002…
∆List(L₃)/∆List(
L₁)
{.004 .006 .016…
■
```

Tip: Be careful about assuming too much from a few table values.

Warning about Linear Looking Graphs

A graph may look linear, but you should not automatically assume that it is linear. Fooled by the local appearance of flatness, for years nearly everyone assumed that the world was flat. In the following example a function is defined whose graph looks linear in a small time frame, but the function is not linear.

[22] The function $P = 67.38(1.026)^t$ models the population of Mexico in the early 1980s. Here P is in millions, and t is the number of years since 1980.

First translate variables and write the formula in Y₁. Then set the viewing WINDOW to $0 \le x \le 6$ (Xscl=1) and $0 \le y \le 100$ (Yscl=50). TRACE shows the equation and graph.

Reset the viewing WINDOW to $0 \le x \le 60$ (Xscl=10) and $0 \le y \le 300$ (Yscl=100). Use TRACE to show the equation and graph. Now we see that over a longer time period the graph is not linear.

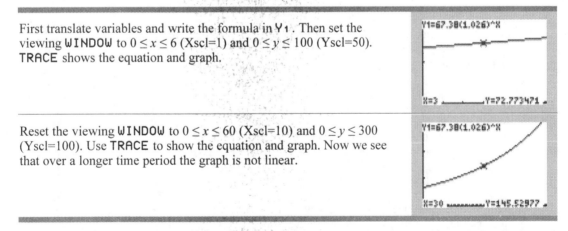

Tip: Table values are more reliable here. They would show P is not linear over $0 \le x \le 6$.

1.4 Graphs and Formulas for Linear Functions

In this section we will look at some features of the graphing calculator that aid in getting a good viewing window for a graph.

Soda and Chips Example

[28]   We have $24 to spend on soda and chips for a party. A six-pack of soda costs $3 and a bag of chips costs $2. The number of six-packs we can afford, y, is a function of x, the number of bags of chips we decide to buy. Find an equation relating x and y; graph the equation and interpret the intercepts and other x, y values.

If we buy x bags of chips we have spent $2x$ dollars, if we buy y six-packs of soda we have spent $3y$ dollars, and the total is $24. An equation relating x and y is:

$$2x + 3y = 24 .$$

However, this form of a linear equation cannot be entered into the calculator. We solve for y and find

$$y = 8 - \frac{2}{3}x .$$

In all the previous examples, the viewing window has simply been given. The decision on a viewing window is the most difficult part of graphing.

Finding a Good Viewing Window

After entering Y₁=8-(2/3)X, press the ZOOM key to show the ZOOM menu shown. Select option 6:ZStandard to set the viewing window to $-10 \leq x \leq 10$ (Xscl=1) and $-10 \leq y \leq 10$ (Yscl=1). This is a good "first take" window for functions involving small numbers.

We see that the y-intercept is 8. This makes sense because if we buy 8 six-packs, we have no money left for chips. If we buy all chips, we know we can buy 12 bags. (This point is not in the window.) Also, in this situation, negative values are meaningless. So our viewing window can be improved.

After the above reasoning, we set the viewing window values as shown. Note that we have set the Ymax value to be greater than the y-intercept so that the formula at the top of the screen does not obscure the intercept. (You could also set Ymin= ⁻2 to leave room for values at the bottom.)

Now TRACE gives a good viewing window. Viewing windows are a matter of taste and may also depend on what you want to do with them. For example, if you want to use TRACE and see how many six-packs to buy for two or three bags, then this window is lousy. (Try tracing to exactly $x = 2$ or $x = 3$.)

There are ZOOM options that reset screens in various ways. The 8:ZInteger will reset the window so that the X values will be integers when traced. (Just what we need in this case.) Beware, this is a two-step process. After pressing 8, the cross-hair cursor will blink on the screen for you to select a new window center. It hardly matters in this case, so we just press ENTER. Now press TRACE, and integer values of x are shown when tracing. Notice, however, that our viewing window is not as good as the previous one.

Can we have our cake and eat it too? Return to the previous settings with **ZOOM MEMORY 1:ZPrevious** (or reenter them manually.) Press **TRACE 2**. (This shows x and y with $x = 2$.) After these steps the screen should appear as shown.

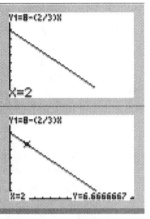

Press **ENTER** and you will see that trace has gone directly to the exact value you wanted.

> *Tip*: For a single value of the function, the above method is quite handy, but if you want to see many values, say from 0 to 12, then a table is the best choice.

ZOOM Options

1:ZBox Resets the window after the box is determined. This allows you to draw a rectangular box on the screen, which will be the new viewing window.

2:Zoom In Resets after centering. Move the cursor to the center of where you want to look closer and press **ENTER** to see a close-up view.

(The scale of zooming in and out is set in **ZOOM MEMORY 4:SetFactors...**).

3:Zoom Out Resets after centering. Move the cursor to the center of where you want to back away from and press **ENTER**.

4:ZDecimal Sets the viewing window to the seemingly bizarre values $-4.7 \le x \le 4.7$ (Xscl=1) and $-3.1 \le y \le 3.1$ (Yscl=1). To view a function close to the origin, this setting has the advantage of tracing x values that are the exact decimal values of -4.7, -4.6,…, 0, 0.1, ..., 4.6, 4.7. In other words, tracing would evaluate 1.5 exactly, not some long decimal close to it.

5:ZSquare Resets the aspect ratio of the window. Since the screen is not square some window settings make circles look like ellipses or perpendicular lines may not appear to have right angles. *Note*: **ZDecimal** is already a square setting.

6:ZStandard Sets the viewing window to the values $-10 \le x \le 10$ (Xscl=1) and $-10 \le y \le 10$ (Yscl=1).

7:ZTrig Sets the window to nearly $-2\pi \le x \le 2\pi$ but it has an exact Xscl=$\pi/2$. This is useful for tracing trigonometric function values. We will use this in Chapter 6.

8:ZInteger Resets, after centering, to trace integer values.

9:ZoomStat Resets after calculating a suitable window for a data list graph. See Section 1.6.

0:ZoomFit Resets using the current **X** values and finds suitable **Ymin** and **Ymax** values.

> *Tip*: The distinction between set and reset in the above list is helpful. Set will give the same window every time, but reset will make an adjustment based on the current window. You can expect a cross-hair cursor to appear for boxing and centering. Normally, center readjustment is unnecessary, and pressing **ENTER** moves you along quickly.

Viewing Windows for Applications

In closing this section, we note that the set zoom keys (4, 6, and 7) are often good for the kind of formulas that need to be graphed in abstract mathematics, such as $y = 2x + 3$ or $y = x^2$. But in actual applications the input and output values are often large or restricted (or both), and the zoom window is inappropriate. In most applications the input values, the X settings, are known.

[22]   For example, in Section 1.3 we saw the population of Mexico function, $P = 67.38(1.026)^t$, and we knew the input values. We wanted to look in the short-run, 0 to 6 years, and then in the long-run, 0 to 60 years. It was less clear what y values should be used for the viewing window. One simple way to find them is first to do a table and see what kinds of output values need to be accommodated. A quicker way to get y value window settings is with the 0:ZoomFit.

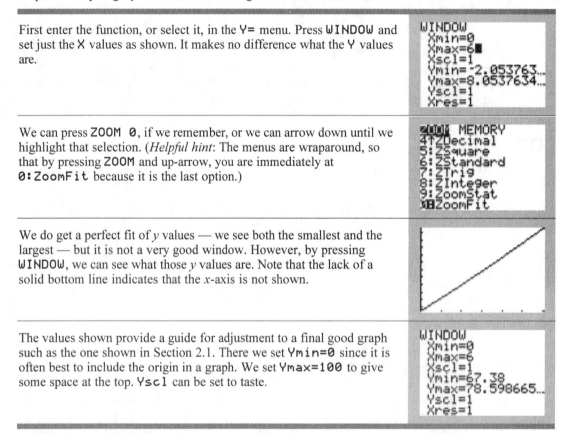

First enter the function, or select it, in the Y= menu. Press WINDOW and set just the X values as shown. It makes no difference what the Y values are.

We can press ZOOM 0, if we remember, or we can arrow down until we highlight that selection. (*Helpful hint*: The menus are wraparound, so that by pressing ZOOM and up-arrow, you are immediately at 0:ZoomFit because it is the last option.)

We do get a perfect fit of y values — we see both the smallest and the largest — but it is not a very good window. However, by pressing WINDOW, we can see what those y values are. Note that the lack of a solid bottom line indicates that the x-axis is not shown.

The values shown provide a guide for adjustment to a final good graph such as the one shown in Section 2.1. There we set Ymin=0 since it is often best to include the origin in a graph. We set Ymax=100 to give some space at the top. Yscl can be set to taste.

Now graph the population function again for $0 \le x \le 60$ using the ZoomFit feature. As in Section 1.3, this graph tells us that the function is not linear.

1.5 Multiple Linear Functions

In this section we see how to graph multiple functions. The main emphasis will be on finding the intersection point of two lines.

The Effect of Parameters:

In the formula $y = b + mx$, we recognize x as the input and y as the output, but what are b and m? These are some constant values that are set before we work with the equation. We call these *parameters*. In the following example, we will see that b tells us where the line crosses the y-axis and m tells us about the inclination of the line.

In the $Y=$ screen we define three functions $Y_1 = 1 + X$, $Y_2 = 2 + X$, and $Y_3 = 3 + X$.

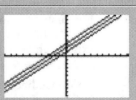

Press ZOOM ZStandard to see the three functions graphed in the same window.

Using graph styles

Multiple function graphs may need a graph style to make them more distinguishable. By arrowing to the far left of an equation definition and pressing **ENTER** you can change to the thick ＼ graph style for Y_2. The series of seven styles are $\{ \diagdown, \searrow, \blacktriangledown, \blacktriangle, \diamond, 0, \because \}$. These are selected by repeatedly pressing **ENTER** to cycle through the options. See the *TI-Guidebook* for more details.

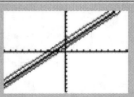

The thick style is assigned to Y_2 to make it distinguishable from the other two lines.

Tip: Using the dotted style will not always show differently from the standard line style.

Tip: Clearing a function definition automatically resets the graph style to the standard line.

Using a List in a definition: a parameter trick

This is a cheap magic trick, but it's fun. In the $Y=$ screen we define the function $Y_1 = \{1, 2, 3\} + X$ and use a ZStandard window. Hopefully, you see through the trick. This single formula defines three functions: $y = 1 + x$, $y = 2 + x$, and $y = 3 + x$.

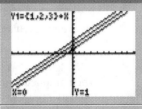

Turn off (deselect) Y_1 and define the function $Y_2 = \{1, 2, 3\}X$. Here again use a ZStandard window. You should be able to explain this screen based on the previous screen.

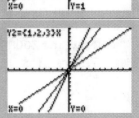

Tip: There is a disadvantage in using the above magic trick. If you use TRACE, the cursor can still jump from graph to graph, but the label does not change so you may not know which function formula you are actually on.

Horizontal and Vertical Lines

[37] It is simple to show a horizontal line: just set the Y variable to some number. However, a vertical line is not a function and thus cannot be defined via the Y= menu. You can draw vertical lines on an existing graph as shown below.

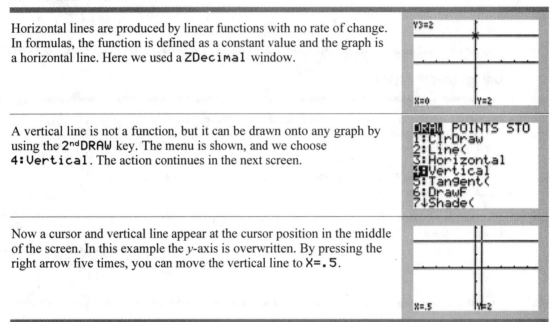

Horizontal lines are produced by linear functions with no rate of change. In formulas, the function is defined as a constant value and the graph is a horizontal line. Here we used a ZDecimal window.

A vertical line is not a function, but it can be drawn onto any graph by using the 2nd DRAW key. The menu is shown, and we choose 4:Vertical. The action continues in the next screen.

Now a cursor and vertical line appear at the cursor position in the middle of the screen. In this example the y-axis is overwritten. By pressing the right arrow five times, you can move the vertical line to X=.5.

Parallel and Perpendicular

Parallel lines have the same slope (rate of change). All three functions $y = 1+x$, $y = 2+x$, and $y = 3+x$ have a slope of 1, and so they are parallel in the first graphing sequence of this section. Perpendicular lines have slopes that are the negative reciprocals of each other, but depending upon the viewing window the TI may give you a distorted view.

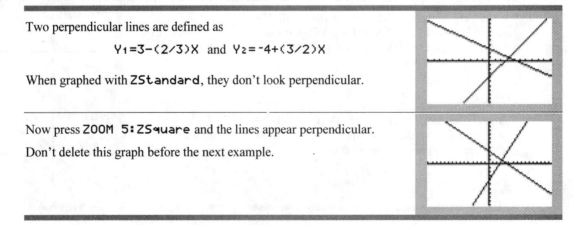

Two perpendicular lines are defined as

Y₁=3-(2/3)X and Y₂=-4+(3/2)X

When graphed with ZStandard, they don't look perpendicular.

Now press ZOOM 5:ZSquare and the lines appear perpendicular.

Don't delete this graph before the next example.

Finding Points of Intersection

It is a common need to find the intersection of graphs. We start with finding the intersection point of the two lines in the last example and then go on to find multiple intersections for an application.

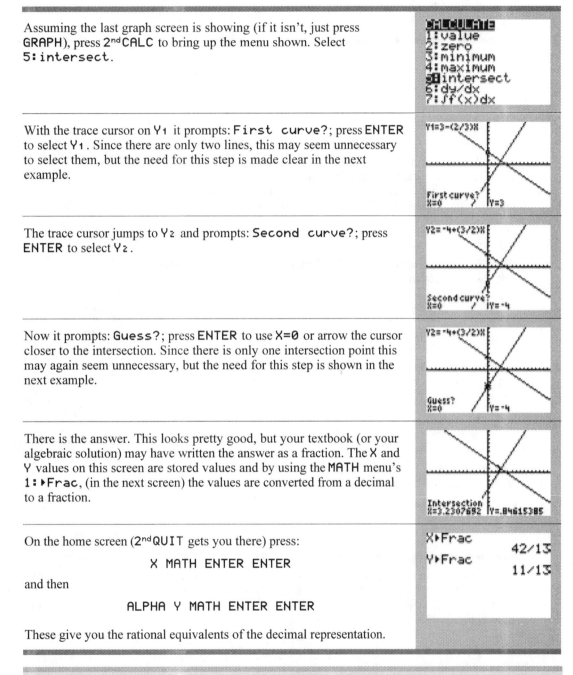

Assuming the last graph screen is showing (if it isn't, just press GRAPH), press 2nd CALC to bring up the menu shown. Select 5:intersect.

With the trace cursor on Y1 it prompts: First curve?; press ENTER to select Y1. Since there are only two lines, this may seem unnecessary to select them, but the need for this step is made clear in the next example.

The trace cursor jumps to Y2 and prompts: Second curve?; press ENTER to select Y2.

Now it prompts: Guess?; press ENTER to use X=0 or arrow the cursor closer to the intersection. Since there is only one intersection point this may again seem unnecessary, but the need for this step is shown in the next example.

There is the answer. This looks pretty good, but your textbook (or your algebraic solution) may have written the answer as a fraction. The X and Y values on this screen are stored values and by using the MATH menu's 1:▶Frac, (in the next screen) the values are converted from a decimal to a fraction.

On the home screen (2nd QUIT gets you there) press:

        X MATH ENTER ENTER

and then

        ALPHA Y MATH ENTER ENTER

These give you the rational equivalents of the decimal representation.

Tip: When using ▶Frac, if a decimal value cannot be converted, or the resulting denominator has more than three digits, then the decimal value is shown.

Which car agency is cheaper?

[35]   The cost, C, in dollars of renting a car for a day from three different rental agencies and driving the car d miles is given by the following three functions:

$$C_1 = 50 + 0.10d \qquad C_2 = 30 + 0.20d \qquad C_3 = 0.50d.$$

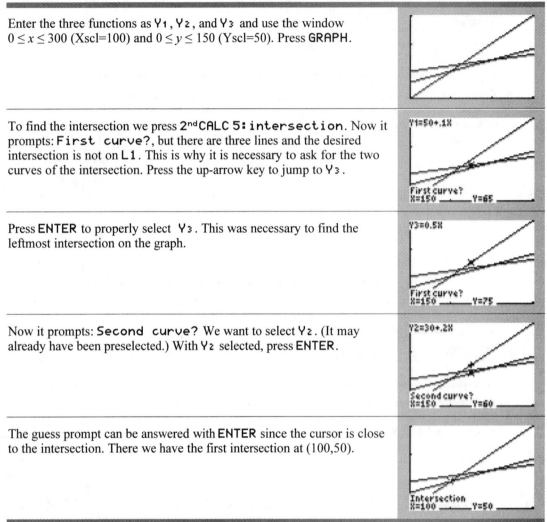

Enter the three functions as Y_1, Y_2, and Y_3 and use the window $0 \leq x \leq 300$ (Xscl=100) and $0 \leq y \leq 150$ (Yscl=50). Press **GRAPH**.

To find the intersection we press **2nd CALC 5: intersection**. Now it prompts: **First curve?**, but there are three lines and the desired intersection is not on **L1**. This is why it is necessary to ask for the two curves of the intersection. Press the up-arrow key to jump to Y_3.

Press **ENTER** to properly select Y_3. This was necessary to find the leftmost intersection on the graph.

Now it prompts: **Second curve?** We want to select Y_2. (It may already have been preselected.) With Y_2 selected, press **ENTER**.

The guess prompt can be answered with **ENTER** since the cursor is close to the intersection. There we have the first intersection at (100,50).

The other intersections can be found at (125, 62.5) and (200, 70). From this we conclude that agency three is cheapest up to 100 miles, agency two is cheapest between 100 and 200 miles, and agency one is cheapest for more than 200 miles. As it turns out, we didn't actually need to find the intersection at 125.

1.6 Fitting Linear Functions to Data

Laboratory Data: Viscosity of Motor Oil

[42]   The viscosity of a liquid (its resistance to flow) depends on the liquid's temperature. Viscosity of motor oil is a measure of its effectiveness as a lubricant in the engine of a car. Thus, the effect of the engine temperature is an important determinant of motor oil performance. See the following table for example data.

T, temperature (°F)	v, viscosity (lbs·sec/in^2)
160	28
170	26
180	24
190	21
200	16
210	13
220	11
230	9

Find a formula that can be used to approximately predict the viscosity of the oil at a given temperature.

Graphing a Scatter Plot

Press **STAT ENTER** to enter the list editor. If necessary clear previous data in the list. Enter the values from the data table. Note that not all the values can show on the screen at one time.

Press 2nd **STAT PLOT** to show a menu of the three special plots called **Plot1**, **Plot2** and **Plot3**. In this screen we see they are all turned **Off**. We need to set the first one to **On**.

We select the first plot by pressing **1**, and are shown a setup screen that allows us to make all the graph settings (except for the window). The settings are changed by arrowing to a choice and pressing **ENTER**. First turn the plot **On**. The **Type** icon shown is a scatter plot. The **Xlist** and **Ylist** are set by pasting in **L1** and **L2**. The **Mark** for each point is a square.

Press Y= to return to the function definition screen. On this screen you will now see **Plot1** highlighted. Turn off any functions that are unrelated to this plot.

```
Plot1 Plot2 Plot3
\Y1=50+.1X
\Y2=30+.2X
\Y3=0.5X
\Y4=
\Y5=
\Y6=
\Y7=
```

We could look at the data and figure out a good viewing window but the **ZOOM** menu has a special setting **ZoomStat** that does this automatically.

```
ZOOM MEMORY
4↑ZDecimal
5:ZSquare
6:ZStandard
7:ZTrig
8:ZInteger
9:ZoomStat
0:ZoomFit
```

Now we see the data points graphed as little square marks. However there are no axes and it may be necessary to press **WINDOW** to see what the window settings are. (You can also press **TRACE** and jump from point to point.)

We reset the window slightly to see the horizontal axis. The new window is used as we proceed to the next stage where we draw an approximating line on the graph.

```
WINDOW
 Xmin=160
 Xmax=250
 Xscl=10
 Ymin=0
 Ymax=35
 Yscl=5
 Xres=1
```

Tip: After a **Plot** is setup properly, you can turn it on and off from the top of the Y= menu.

Regression Curves

A regression curve is a function that approximates the data. In this case we see that the data values approximate those of a linear function. The approximating line is called a regression line.

In the coming section, your screen may not appear exactly like the one shown. To avoid dealing with this in the middle of the example we now make a settings change. Start with a clean home screen and use 2nd**CATALOG** to paste in **DiagnosticsOn**. Press **ENTER**, and **Done** appears. Once this is done, this setting stays on unless it is deliberately reset to be off.

Finding a regression line

This screen is our scatter plot using the window from the previous plot.

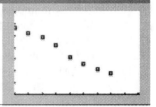

Press **STAT** and arrow to the **CALC** menu. Make the choice **4:LinReg(ax+b)**. Choices **4** and after are the different kinds of regression curves that you can use. We use other regression curves in Section 3.3 and Section 9.7.	EDIT **CALC** TESTS 1:1-Var Stats 2:2-Var Stats 3:Med-Med **4▮**LinReg(ax+b) 5:QuadReg 6:CubicReg 7↓QuartReg
After pasting **LinReg(ax+b)** on a clean home screen, we specify the two lists and (optionally) a **Y** to write the regression equation into. We are using **L1** and **L2** and pasting the equation into **Y1**. Press **ENTER** to see the screen in the next frame.	LinReg(ax+b) L₁, L₂,Y₁
This gives the equation of the regression line, but it also has other information. The **r** is called the correlation coefficient, and it tells you the closeness of the approximation equation. See your text for more information.	LinReg y=ax+b a=-.2928571429 b=75.60714286 r²=.9841920375 r=-.9920645329 ▮
We press **TRACE** to see the scatter plot and the regression line. Press the up arrow to see that at the top of the screen the regression equation has been pasted to **Y1** with full accuracy.	Y1=-.29285714285714X+75... X=205 Y=15.571429

Interpolation and Extrapolation

So what is the use of a regression line? It is a compact formula that can be used to approximate the data values. When it approximates a value between known values it is said to be an interpolation. If the value is before or after known values, then it is called an extrapolation.

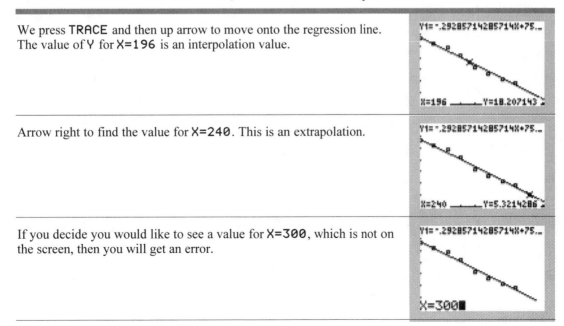

We press **TRACE** and then up arrow to move onto the regression line. The value of **Y** for **X=196** is an interpolation value.	Y1=-.29285714285714X+75... X=196 Y=18.207143
Arrow right to find the value for **X=240**. This is an extrapolation.	Y1=-.29285714285714X+75... X=240 Y=5.3214286
If you decide you would like to see a value for **X=300**, which is not on the screen, then you will get an error.	Y1=-.29285714285714X+75... X=300▮

To remedy this error, you can expand the window or use just the TABLE feature.

```
ERR:INVALID
1█Quit
2:Goto
```

CHAPTER TWO

FUNCTIONS, QUADRATICS, AND CONCAVITY

2.1 Input and Output

For a function, each input has a unique output. For functions defined by a formula we now review how to find this output for a given input. This is called *evaluating the function*. Next we reverse the process: for a given output we find an associated input value. In the equation $y = f(x)$, this means that we know y and want to find x. This reverse process is called *solving the equation*.

Evaluating a Function Using a Formula

Recall from Section 1.2 that when a function is defined using the notation $y = f(x) = x^2$, then $f(10) = 10^2$ and we write $f(10) = 100$. In words: "We are evaluating the function at 10." For example, to evaluate the function f using the TI, define Y₁=X^2 using the Y= menu (or from the home screen), and then enter Y₁(10) on the home screen. The mathematical notation $f(10) = 100$ is translated to the TI notation Y₁(10) with answer 100.

Numerical vs. algebraic evaluation

[75] Let's work on a more challenging function

$$g(x) = \frac{x^2 + 1}{5 + x}.$$

Evaluate the following expressions

(a) $g(3)$ (b) $g(-1)$ (c) $g(a)$ (d) $g(a - 2)$.

First translate variables and write the formula in Y₁. Note the necessity of parentheses to force the correct order of operations in the fraction. Use 2ⁿᵈQUIT to return to the home screen.	Plot1 Plot2 Plot3 \Y₁◘(X^2+1)/(5+X) \Y₂= \Y₃=

We can evaluate the function Y₁ at 3 and −1. (In Section 1.2 we learned how to paste the Y₁ to the home screen by using the VARS key.) After finding Y₁(3) you can use 2ⁿᵈENTRY to save a trip back to the VARS menu when finding the second function value.	Y₁(3)
	           1.25
	Y₁(-1)
	           .5

To find $g(a)$, we might hope that $Y_1(A)$ would give an algebraic answer in terms of A, but the TI operates only at the numerical level. Thus the expression $Y_1(A)$ will give the value of the function using the current value of A. Look at your current value of A (which may vary from the one shown) and then evaluate the function to verify this.

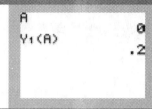

The more advanced calculators like the TI-89 and TI-92 have symbolic capabilities that actually evaluate part (c) and part (d) algebraically. By using algebra or a TI-89, you can confirm:

$$g(a) = \frac{a^2+1}{5+a} \quad \text{and} \quad g(a-2) = \frac{a^2-4a+5}{3+a}.$$

> *Tip:* Evaluating a function from the graph screen is possible by pressing 2nd CALC 1:value and giving an X value between the Xmin and Xmax of the viewing window.

Finding Input Values from Output: Solving Equations

[76]   Suppose $y = \dfrac{1}{\sqrt{x-4}}$.

(a) Find an x-value that results in $y = 2$.
(b) Is there an x-value that results in $y = -2$?

Using Solver

Part (a) is asking us to solve the equation

$$2 = \frac{1}{\sqrt{x-4}}.$$

We could do this algebraically by hand, but the TI does the job as shown below. But first, to get our bearings, we graph the function.

Graph the function in a ZStandard window. Use TRACE and arrow over as shown. (In the next section we talk later about the strangeness of this graph.) The cursor point has a value of y close to 2, so that the values of x we are looking for is about 4. Thus, we have approximately solved for x.

Y1=1/√(X-4)
X=4.2553192 Y=1.979057

We now find a more accurate solution. Press MATH 0 as shown to begin the Solver.

```
MATH NUM CPX PRB
4↑³√(
5:×√
6:fMin(
7:fMax(
8:nDeriv(
9:fnInt(
0Solver…
```

An equation line appears on which to complete the eqn:0= prompt for the equation. Notice that we have to transform the equation by moving all the terms to the right. (If some previous equation is shown, use CLEAR before starting the new equation entry.) Press ENTER or the down arrow when the equation is entered.

```
EQUATION SOLVER
eqn:0=2-1/√(X-4)
```

The X value from the graph is shown as a guess. Now press ALPHA SOLVE. (SOLVE is above the ENTER key.) This command says: solve the equation with the guess shown. The bound= line can be used to restrict the x values it looks for, but we ignore that line for now.

```
2-1/√(X-4)=0
 X=4.25531915■
 bound={-1E99,1...
```

Once solved, the solution is shown with a black square block to the left of the freshly calculated variables. Do not trust a Solver screen until you see this symbol. The additional line left-rt=0 indicates that this solution is exact. When this line is not zero it alerts you that the solution is an approximation by giving the difference between the left and right side of the equation.

```
2-1/√(X-4)=0
■X=4.25
 bound={-1E99,1...
■left-rt=0
```

The difference between Solver and solve

Solver is a new and improved version of the solve command from the TI-82. The solve command of the TI-83 is not on any menu but it is still available in the CATALOG. The syntax for solve is:

$$solve(expression, variable, guess).$$

The result is a value of the *variable* that makes the *expression* zero by starting to search for that value at *guess*. The previous Solver equation solution is redone below using solve.

On the home screen, paste solve(and complete the command as shown. We have used a guess of 5 because we know from the graph that the function is undefined for $x \leq 4$. A common blind guess of zero is frequently used in the solve command, but in this case it would cause an error. Such an error is shown in the next subsection.

```
solve(2-1/√(X-4)
,X,5)
          4.25
```

Tip: Solver and solve allow you to enter expressions containing the Y= variables. For example, solve(2-Y₁,X,5) is the same as the above screen's command, when Y₁=1/√(X-4).

Solve and Solver error warnings

Because they use numeric techniques, solve and Solver can be in some cases unreliable. In the first example below they fail to find a solution that actually exists. The second example explains the error message when you try to find a solution that doesn't exist. This will answer our original problem's part (b): Is there an x-value that results in $y = -2$?

In Solver, first check to see that the proper equation is entered. (This step not shown, but the proper equation is shown on the top line of the screen.). Now we attempt to re-create the known solution with a guess of X=0. The surprise is that we get an error message.

```
2-1/√(X-4)=0
 X=0■
 bound={-1E99,1...
```

This error message should not be interpreted as telling us that there is no real answer. We already know there is one! By starting with $x = 0$, the value $√(-4)$ is encountered which is nonreal.

```
ERR:NONREAL ANS
1■Quit
2:Goto
```

Now we attempt to answer part (b) by using $y = -2$. In this case the function is always positive and thus could never be equal to -2. There is no x value that can be found, but the error message doesn't explicitly say that, as shown on the next screen.

```
-2-1/√(X-4)=0
X=4.25■
bound=(-1ε99,1...
```

The `Solver` method in finding a zero is to find one positive and one negative value of the difference, then it squeezes in on the zero that is trapped between. In cases where all the values of the function are positive it eventually quits and says: `NO SIGN CHNG`, (no sign change.)

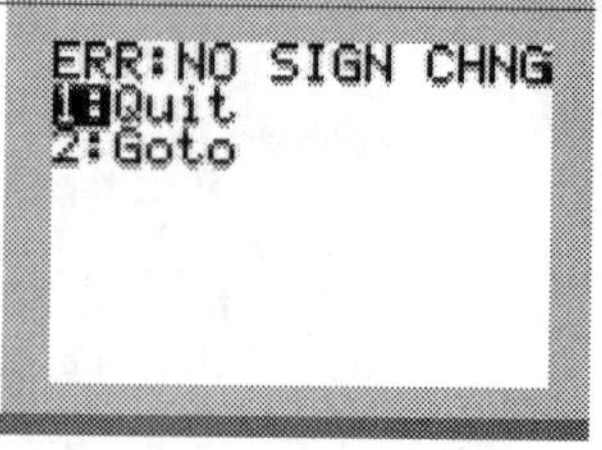

```
ERR:NO SIGN CHNG
1█Quit
2:Goto
```

Tip: Using a quick graph in the beginning will inform you if there will be solutions and give you good approximations that avoid error messages.

2.2 Domain and Range

A good viewing window for a function's graph provides broad hints about the domain and range of the function. This section focuses on finding good viewing windows and interpreting the results about the domain and range.

Sunflower Growth

[88]   A sunflower plant is measured every day t, for $t \geq 0$. The height, $h(t)$ centimeters, of the plant can be modeled by using the function

$$h(t) = \frac{260}{1 + 24(0.9)^t}.$$

What is the domain of this function? What is the range?

A common-sense estimate of the growing season is 90 days. We are given the restriction $t \geq 0$; thus we can set the window as $0 \leq t \leq 90$. In `WINDOW`, set `Xmin=0` and `Xmax=90`. Use `ZoomFit` to give a good first window. If you do not get this graph, check your function definition for correct parentheses.

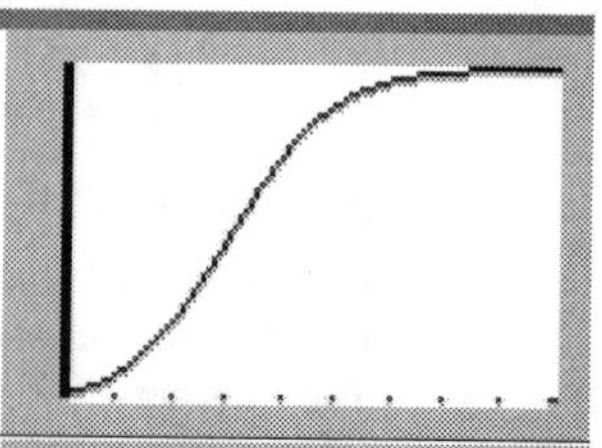

The axis values are not shown, so press `WINDOW` to see the range of values. Our answer would be that for a domain of $0 \leq t \leq 90$ the range is $10.4 \leq h(t) \leq 259.53$. The problem said that the model could be used for $t \geq 0$. Although it may not make sense in this model, let's see what happens to the $h(t)$ values, as t gets larger and larger.

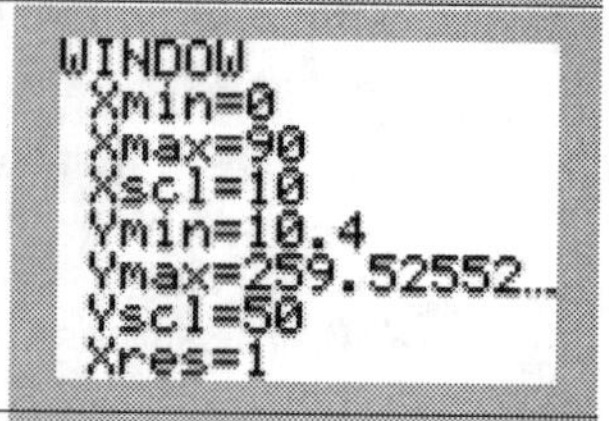

```
WINDOW
Xmin=0
Xmax=90
Xscl=10
Ymin=10.4
Ymax=259.52552...
Yscl=50
Xres=1
```

Use `TRACE` to look at values of `Y1` for `X` larger than 90. Hold down the right-arrow key and the cursor moves to the right until it reaches 90. At that point it *pans* (shifts) over and show values greater than 90. In the screen shown, the cursor has disappeared, but the values can be read at the bottom of the screen.

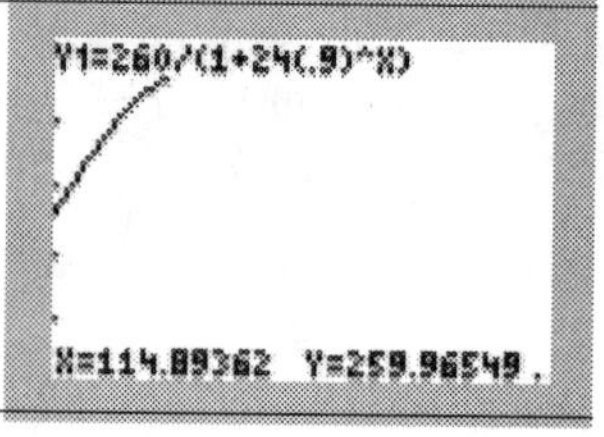

```
Y1=260/(1+24(.9)^X)




X=114.89362  Y=259.96549
```

The **WINDOW** settings $0 \leq x \leq 90$ and $-50 \leq y \leq 300$ will produce the screen shown. (Extra room is given so that tracing information does not obscure the top or bottom of the screen.) Now by panning to the right, you can convince yourself that the **Y** values get closer and closer to 260.

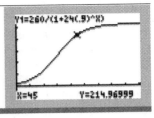

In summary, if the domain is $t \geq 0$, then the range is $10.4 \leq h(t) < 260$ for this model. The mathematical domain of this function is not restricted by reality, it is all real numbers. This will be verified when we discuss exponential function in the next chapter.

Tip: In general, when setting a **WINDOW** make the **Y** settings generous to accommodate for the top and bottom information in **TRACE**.

Undefined Points

[89]   A single undefined point in the domain of a function can often be hard to see on a graph; however, undefined intervals are easier to spot. We return to our strange graph from Section 2.1,

$$y = \frac{1}{\sqrt{x - 4}}.$$

Because we can only take the square root of a positive number, we see that this function is undefined for $x < 4$. In addition, if $x = 4$, then we have a zero in the denominator which is also not defined. Thus our domain is $x > 4$. We now see how the TI communicates these conclusions.

Undefined domain values

Using **ZStandard** and **TRACE**, we see our graph. A very strong hint that the function is undefined is the fact that at **X=0** there is no **Y** value given. Repeated pressing of the right-arrow key shows that **Y** remains undefined until you reach points with x greater than 4.

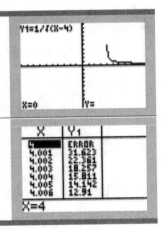

We check this out by using a table. In a table, an undefined domain value has **ERROR** as the **Y** values. (See $x = 4$.) The table increment was set as 0.001 so that we could see the numeric behavior close to 4. The values are getting larger and larger as we get close to 4.

Undefined points on the domain and range

[90]   Suppose we have the function $y = 2 + \dfrac{1}{x}$. We see right away that the denominator is x, so that it is not defined for $x = 0$. Algebraically, we can also show that the range is all real values except $y = 2$. The graph will give us hints that we are correct.

We first graph the function using ZStandard and TRACE; then we interpret the screen. The domain does not include X=0 since there is no Y= value given. We next use a table to decide if the range includes large positive/negative values since it disappears into the vertical axis. There may also be a break in range values at $y = 2$.

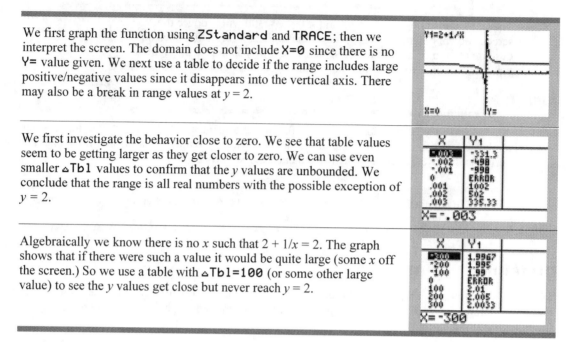

We first investigate the behavior close to zero. We see that table values seem to be getting larger as they get closer to zero. We can use even smaller ⌂Tbl values to confirm that the y values are unbounded. We conclude that the range is all real numbers with the possible exception of $y = 2$.

Algebraically we know there is no x such that $2 + 1/x = 2$. The graph shows that if there were such a value it would be quite large (some x off the screen.) So we use a table with ⌂Tbl=100 (or some other large value) to see the y values get close but never reach $y = 2$.

We conclude that this function has a domain of all real numbers except $x = 0$ and a range of all real numbers except $y = 2$. The graph and/or table did not specifically state that this is the case. To be absolutely sure, we must make an algebraic argument. The TI can be used first to make a good guess at the domain and range. Or, after having found the domain and range algebraically, the calculator can serve as a confirmation.

2.3 Piecewise Defined Functions

In real life there are many piecewise defined functions, one example is the rule used to calculate our income tax. This tax function is published as a table that gives the tax due for an income bracket. A piecewise defined function is a collection of two or more rules, and the particular rule used for output, depends on the value of the input. As another example, if you are boiling water a quart of water on your stove, then the temperature of the water depends upon the number of minutes since you started heating it. Thus, there is a formula to model the temperature before reaching the boiling point, and a second formula that goes into effect for the time after the boiling point has been reached. (Remember: After the boiling point is reached the temperature is constant.)

The Absolute Value Function

[74]   A piecewise definition for the absolute value often has the form

$$f(x) = \begin{cases} x & \text{for} \quad x \geq 0 \\ -x & \text{for} \quad x < 0. \end{cases}$$

One rule is for x greater than or equal to zero, and another rule is for x less than zero.

Let's graph this function. From the Y= menu, use 2ⁿᵈCATALOG to paste in the abs function.

The parentheses may seem useless in this case, but in most cases they are needed to specify the expression that will be used as input for the function.

To see a graph, we use ZDecimal to give a good window.

Here we see two linear functions graphed. One with slope +1 and the other with slope −1. It would be wrong to define Y₁=X and Y₂=−X and then graph. There is a domain restriction for each formula.

Tip: A common error in graphing piecewise defined functions is to graph all the formulas over the complete domain.

Logical Expressions

The domain is usually designated by intervals, such as $x > 0$. If $x > 0$ is wrapped in parentheses it is called a logical expression and the calculator can check to see if it is True or False. A True expression has a value of 1; a False expression has a value of 0. To define a piecewise function, we introduce a little trick that use logical expressions to restrict each formula by including its value only on its particular interval. For example, the absolute value function could be written as

$$f(x) = -x\,(x < 0) + x\,(x \geq 0).$$

This works because when the logical is False then that part of the sum will be zero, but when it is True then we have the formula's value. Let's do a more difficult example. Graph the function

$$g(x) = \begin{cases} x+1 & \text{for} \quad x \leq 2 \\ 1 & \text{for} \quad x > 2 \end{cases}$$

For this example, it will be helpful for you to read the first three steps below before starting to enter anything in your calculator.

The definition of g can be translated into

$$g(x) = (x+1)(x \le 2) + (1)(x > 2),$$

where the two inequalities act as logical expressions. We start to enter the function in Y1 but come to a point where we now need the logical symbol $\le$.

Logical symbols are listed in the 2ⁿᵈTEST menu. Here we paste the $\le$ symbol from the 2ⁿᵈTEST menu. Continue to enter the whole expression.

Here is how it works: If $x \le 2$, the expression $\langle X \le 2 \rangle$ has a value of 1 but $\langle X > 2 \rangle$ has a value of 0. This means that Y1 is Y1=$\langle X+1 \rangle \langle 1 \rangle + \langle 1 \rangle \langle 0 \rangle$, which is $y = x + 1$. But if $x > 2$, the expression $\langle X \le 2 \rangle$ has a value of 0, but $\langle X > 2 \rangle$ has a value of 1. Now Y1=$\langle X+1 \rangle \langle 0 \rangle + \langle 1 \rangle \langle 1 \rangle$, which is $y = 1$.

Graph using ZDecimal and get a little surprise. It has connected the values on either side of $x = 2$. When the TI graphs a function, it evaluates 95 x-values to plot 95 points. These points are normally connected. In this case the connection on either side of $x = 2$ should not be made.

To change this we go into the MODE menu, and we arrow down and over to Dot. Press ENTER to make the selection. Press 2ⁿᵈQUIT to leave the MODE screen. For most piecewise defined functions, this is the best setting.

Graph again to see the difference.

A Table to Draw Open/Closed Circles on Graphs

Note that if you copy the previous graph as part of an assignment, your instructor may want you to include an open circle for endpoints not included and a closed circle for included endpoints. Most textbook graphics use this convention. See page 72 of *Functions Modeling Change*.

Use **TABLE** with an initial value of 2 and an increment of 0.001. Then up arrow to see some values before and after 2. We see that the value of the function at 2 is 3, $g(2) = 3$. Therefore, the line $y = x + 1$ should have a closed circle as an endpoint at $x = 2$. When x-values are above 2 the function value is 1, so the horizontal graph $y = 1$ should have an open circle for the endpoint at $x = 2$.

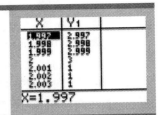

2.4 Inverses

The TI-83 does not find the inverse of a function. However, using **Solver** it allows you to easily find values of a function's inverse.

Inverse Functions

[78]   There is a special mathematical notation for the inverse value of a function. For example, the cricket chirp rate formula and its inverse would be written:

$$\text{if } T = f(R), \text{ then } R = f^{-1}(T).$$

This inverse notation cannot be used on the TI-83; however, **Solver** makes it somewhat unnecessary. We now see how to evaluate $f^{-1}(T)$.

Tip: Solver can have many variables, and it solves for any *one* variable using the values set for all the other variables.

Solver in greater glory

In **Solver** the original function formula can be entered as a difference $0 = T - f(R)$. Notice that we did not need to translate to **X** and **Y** variables; however, we did need to use the **ALPHA** key to enter **T** and **R**. (The same work is repeated with **solve** in the next subsection.)

```
EQUATION SOLVER
eqn:0=T-(.25R+40
)■
```

After pressing **ENTER**, all the variables are listed. We now fill in the ones we know (in this example **T=50**), and position the cursor on the line of the variable that we want to know (in this example **R**). To solve, press **ALPHA SOLVE**.

```
T-(.25R+40)=0
T=50
R=■
bound={-1ε99,1...
left-rt=0
```

We see **R=40**. If $T = f(R)$, then **Solver** is evaluating $R = f^{-1}(T)$. In mathematical notation $f^{-1}(50) = 40$. In words this means that at 50 degrees the chirp rate is 40.

```
T-(.25R+40)=0
T=50
•R=40
bound={-1ε99,1...
•left-rt=0
```

The beauty of this system is that we need not get lost in a question about what is the function and what is the inverse function. With the same setup in **Solver** we can calculate either variable in terms of the other. We now calculate the temperature when the chirp rate is 50.

```
T-(.25R+40)=0
T=50
R=50■
bound={-1ε99,1...
```

Arrow up to the line `T=50` and press `ALPHA SOLVE`. We see the newly calculated value `T=52.5`. (The old value, `T=50`, was used as the guess for this calculation.)

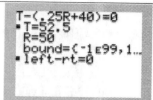

In mathematical notation this result is written as $f(50) = 52.5$, and it means that at the chirp rate of 50, the temperature is 52.5 degrees.

Doing it again with solve

Here is the previous example done on the home screen with `solve`. The colon (:) allows us to do two commands on one line. This is handy when you use the `2nd ENTRY` recall; you can edit the whole line.

2.5 Concavity

Concavity and Rate of Change

For a linear function, the rate of change is constant and its graph has no concavity. If the function's rate of change is increasing we see the graph as concave up; if it is decreasing the graph will be concave down. In general, concavity can be determined from a graph. When the concavity of a graph is hard to see, numerical methods are a better to determine concavity.

[81] The carbon-14 decay function, $Q = 200(0.886)^t$, is not a linear functions; we use a list of rounded values for 0, 5, 10, and 15 thousand years to look at $\Delta Q / \Delta t$.

Enter `Y1=200(.886)^X` and setting the viewing window to $0 \leq x \leq 20$ (Xscl=5) and $-50 \leq y \leq 250$ (Yscl=50), press the `TRACE` key to see the screen shown. This is clearly concave up.

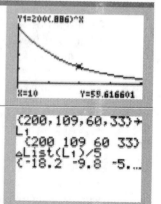

We make a list `L1` of the Q function values (rounded) for $t = 0, 5, 10, 15$. We then find $\Delta Q / \Delta t$ for the three consecutive intervals with $\Delta t = 5$. Since Δt is constant, we do not need to make a difference list for the t values, we just divided by 5. The rate values -18.2, -9.8, and -5.4 are increasing, that is, getting less negative. An increasing rate of change confirms upward concavity.

Tip: The value of the function is decreasing, but the value of the rate of change is increasing.

2.6 Quadratic Functions

The standard form of a quadratic equation is $y = ax^2 + bx + c$, where a, b, c are constants, $a \neq 0$.

Finding Zeros of a Quadratic Function

[85]   Find the zeros of $f(x) = x^2 - x - 6$.

Using a graph to find zeros

Using $Y_1 = X^2 - X - 6$ and Zstandard. We see the graph crosses the *x*-axis at, or close to, −2 and 3. By substitution you can confirm that these are the exact zeros.

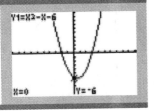

Using solve to find zeros

With Y_1 being as previously defined, we can use solve. In the first calculation we used a guess of 0 giving us our -2 answer. By guessing 3, we find it was exactly the answer.

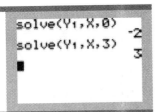

Yet another method of solution is to use Solver. In this case, you can enter the standard quadratic equation 0=A*X^2+B*X+C and then below fill in the known A, B, C and solve for X.

The quadratic formula to find zeros

[86]   The solve (and Solver) method of solution is to start with a guess and use successive calculations to get closer and closer to the correct answer. This gives a very good approximate answer and in some cases an exact answer. The standard quadratic equation has a pair of zeros (solutions) that can be written directly in the exact form:

$$x = \frac{-b \pm \sqrt{b^2 - 4ac}}{2a}.$$

This is not an approximation!

Since the equation is in standard form, we assign (use STO▸) the coefficients to letter names. The colon (:) allows us to make all three assignments as a single command. (The colon symbol is above the decimal point key in ALPHA mode.) When using multiple assignments only the final one, here −6, is shown as output.

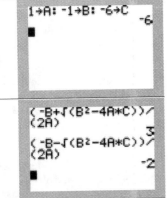

Now enter the famous quadratic formula and calculate the first zero of the function. Use 2nd ENTRY to find the second zero by bringing back the entry and changing the plus sign in front of the square root to a minus sign. For new quadratics you can use deep recall to go back and change the A, B, C line and then reuse the two solution expressions shown here.

The method above, which stores the quadratic formula, is hazardous in that after many non-quadratic formula calculations the formula may "fall off" the deep recall storage. The following optional technique is suggested for students who sleep better at night knowing that their quadratic formula is safely stored. You can write and store the quadratic formula in a simple program.

Optional: Programming Adventure

Press **PRGM** and arrow to **NEW**.	`EXEC EDIT NEW` `1:Create New`
Name the program. Note that the **ALPHA** lock mode is automatically turned on.	`PROGRAM` `Name=QF`
Pressing **ENTER** shows you a program editing screen.	`PROGRAM:QF` `:`

We want our program first to ask for the value of **A**, **B**, and **C**. Press **PRGM** and now, because you are already in program edit mode, it shows a menu screen. Arrow to **I/O** (Input/Output) and select **2:Prompt**. This pastes that word on the program edit screen. (Commands pasted from menus are distinguishable by their lower case letters.)	`CTL I/O EXEC`
	`1:Input`
	`2:Prompt`
	`3:Disp`
	`4:DispGraph`
	`5:DispTable`
	`6:Output(`
	`7↓getKey`

Finish this line by pressing
    ALPHA A , ALPHA B , ALPHA C ENTER.
	`PROGRAM:QF`
	`:Prompt A,B,C`
	`:`

Complete the program as shown. The other lines display the formula for the two calculated zeros. They start with **Disp** from the **PRGM I/O** menu. Press **2nd QUIT** to exit the program edit screen.
 	
Note that parentheses are needed for correct division, that there is a subtraction/negation difference, and that an explicit multiplication sign between **A** and **C** is needed.
	`PROGRAM:QF`
	`:Prompt A,B,C`
	`:Disp (-B+√(B²-4`
	`A*C))/(2A)`
	`:Disp (-B-√(B²-4`
	`A*C))/(2A)`
	`:`

Tip: Programs can be much more complex. The *TI–Guidebook* has a complete chapter that more fully describes the programming capabilities. We started with this example because of its simplicity.

Using a program

From a cleared home screen, press **PRGM**. From the **EXEC** (Execute) menu, select the program you want to use. With the program name on the home screen press **ENTER**.

```
EXEC EDIT NEW
1:QF
```

The program name appears on the top line, and **A=?** prompts you to enter **A**. Enter **A**, then **B**, then **C**. Notice here that the negation key is used to enter a negative number. The subtraction key causes an error. If an error occurs in this program, it is most likely a misuse of the negation and subtraction keys. The programmed formulas contain both negation and subtraction symbols, and they cannot be interchanged.

```
prgmQF
A=?1
B=?-1
C=?-6
```

With luck you were able to see a screen with answers exactly as shown here. If not, go back (**PRGM EDIT**) and edit the program to correct the formulas.

```
prgmQF
A=?1
B=?-1
C=?-6
                -2
              3
            Done
```

Doing it again

[86]   Find the zeros of $h(x) = -\dfrac{1}{2}x^2 - 2$.

Enter **A=-1/2**, **B=0**, and **C=-2**.

```
A=?-1/2
B=?0
C=?-2
```

A simple graph of $h(x) = -\dfrac{1}{2}x^2 - 2$ would show that it does not cross the x-axis. Since there is no real solution, the result is the error message: **NONREAL ANS**. In Chapter 7 we address non-real solutions.

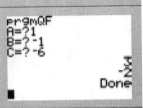

```
ERR:NONREAL ANS
1:Quit
2:Goto
```

Notes:

CHAPTER THREE

EXPONENTIAL FUNCTIONS

3.1 The Family of Exponential Functions

Definition: An *exponential* function has a constant percentage rate of change. In this chapter we look at what this means in terms of tables, graphs, and formulas.

Growing at a Constant Percent Rate

[102] Suppose you are offered a job at a starting salary of $40,000 per year and the company promises annual raises of 6% per year for the first five years after you are hired. Compute the salary for the first few years.

The basic computation each year is:

$$\text{Next salary} = \text{Former salary} + 6\% \text{ of former salary.}$$

Using the TI sequence mode

The TI has a special sequence mode (Seq) that allows you to enter the initial salary and a formula for finding the next value. We now see how this works.

Press **MODE** and change the function mode from Func to Seq in the fourth row of the screen.	Normal Sci Eng Float 0123456789 Radian Degree Func Par Pol Seq
The Y= screen now shows the definitions for three sequence defined functions named u, v, w. The first setting nMin= is preset to zero and defines our starting n. The u(n) is the function definition and u(nMin) is the initial value. In the next frame we see how our sequence function is defined.	Plot1 Plot2 Plot3 nMin=0 \u(n)=█ u(nMin)= \v(n)= v(nMin)= \w(n)= w(nMin)=
Entering the formula requires that you use the 2ⁿᵈu key (above the 7 key). Changing the function mode to Seq redefines the X,T,θ,n key to be the variable n, which is needed in the definition. The u(nMin) can be entered as 40000 without set brackets, but they are automatically inserted. (A more advanced feature allows this value to be a list. See the *TI–Guidebook*, if interested.)	Plot1 Plot2 Plot3 nMin=0 \u(n)=u(n-1)+.06· u(n-1) u(nMin)={40000} \v(n)= v(nMin)= \w(n)=

Before pressing the **TABLE** key to obtain the shown screen, use 2nd **TBLSET** to set **TblStart=0** and Δ**Tbl=1**. This table can be scrolled.

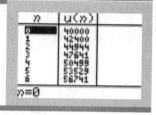

Growth Factor

The above function definition for **u(n)** could also have been written as **u(n)=1.06u(n-1)**. In this form, 1.06 is called the growth factor. Notice that the function, **u(n)**, is only defined for input values 0, 1, 2, and so on. It is called discrete. In truth, the salary function is a piecewise defined function that remains constant throughout the year and then jumps by 6%. We now model the salary function by a formula that has a domain of all real numbers but does not have a sudden jump.

Change the function mode back to **Func**.

Define **Y₁=40000(1.06)^X**.

Set the viewing **WINDOW** to $0 \le x \le 20$ (Xscl=5) and $0 \le y \le 120000$ (Yscl=40000).

Press **TRACE** to show the equation and graph.

Use 2nd **TBLSET** to set **TblStart=0** and Δ**Tbl=1**. This table shows that the functions **Y₁** and **u** have the same value for 0, 1, 2, 3, ... The **Y₁** definition has the mathematical advantage that it is defined for all real numbers and is continuous (without jumps).

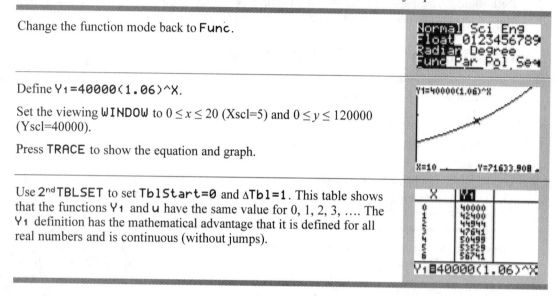

3.2 Comparing Exponential and Linear Functions

We now compare the tables and graphs of exponential and linear functions.

Identifying Linear and Exponential Functions from a Table

[110]   The table below gives values of two functions, f and g. Which function is linear and which is exponential?

x	20	25	30	35	40	45
$f(x)$	30	45	60	75	90	105
$g(x)$	1000	1200	1440	1728	2073.6	2488.32

In this example Δx is constant and it is easy to see that the difference between consecutive values of $f(x)$ is 15. (We could use **ΔList** as in Section 2.1 to identify linear functions.) The difference between consecutive values of $g(x)$ is not constant, but the ratio is constant. We identify $g(x)$ as an exponential function by inspecting the ratios.

Using lists to find quotients

To find consecutive ratios we will make two lists: one is the numerator and one is the denominator. We put all but the last entry for $g(x)$ into list **L1**. (Recall that set brackets and list names require the **2nd** key and that the arrow ($\rightarrow$) is the **STO►** key.) Next edit this list by deleting the first entry, adding the last entry, and storing it as **L2**.	{1000,1200,1440, 1728,2073.6}→L₁ {1000 1200 1440… {1200,1440,1728; 2073.6,2488.32}→ L₂ {1200 1440 1728… ∎
Now **L2/L1** is the list of consecutive ratios. We see that the ratios are all 1.2, which indicates that $g(x)$ is an exponential function. It should be noted that this method of ratio checking requires that Δx be a constant.	{1000 1200 1440… {1200,1440,1728; 2073.6,2488.32}→ L₂ {1200 1440 1728… L₂/L₁ {1.2 1.2 1.2 1.…

Tip: Some students notice that tables with decimal values are likely to be exponential functions, while linear functions are more likely to have whole numbers. While often true, it is not always true; be careful about assuming too much from a few table values.

Exponential Growth Always Outpaces Linear Growth

[114]   Thomas Malthus made gloomy predictions about long-run food supply. He assumed that the population grew exponentially and that food supply grew linearly. For example, if P is the population (in millions) and N is the number of people (in millions) fed from the food supply then we can investigate this model with the two equations

$$P = 2(1.04)^t \text{ and } N = 4 + 0.5t.$$

Enter these functions as **Y1** and **Y2**. Set the viewing **WINDOW** to $0 \le x \le 100$ (Xscl=10) and $0 \le y \le 100$ (Yscl=10). **TRACE** shows the graph. We see that the linear equation outpaces the exponential equation at the 50-year mark, but later the exponential function passes the linear function.	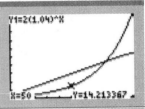
The intersection point is when food shortage will begin. To find this point, you can use the **2ndCALC 5:intersect** feature introduced in Section 1.5. (The intermediate steps are not shown.) We find that the exponential function is greater than the linear function after about 78 years.	

3.3 Graphs of Exponential Functions

In this section we will look at some special features of an exponential graph.

The Exponential Family

[118]   Our basic exponential functions have the form $Q = ab^t$, where a and b are parameters. You can graph a related group of functions to see the effect of changing a parameter. Recall that while it is possible to write each function as its own Y= formula, the set notation can be used to simplify the process. Although the screens are not shown here, you should separately graph each of the following:

$$Y_1=\{150,100,50\}*1.2^X \quad \text{and} \quad Y_2=50*\{1.4,1.2,.8,.6\}^X$$

Horizontal Asymptotes

[119]   The horizontal asymptote of a function is a line $y = k$, such that as $x \to \infty$ or $x \to -\infty$ then $f(x) \to k$. To check this numerically, you can use large values of x in a table. Or you can graph with a very wide window. Exponential functions of the form $Q = ab^t$ have a horizontal asymptote of $y = 0$.

After entering $Y_1=200(.9)^X$ and setting the viewing window to $0 \le x \le 60$ (Xscl=15) and $-50 \le y \le 250$ (Yscl=50), press the **TRACE** key to see the screen shown. The graph falls toward the x-axis, and at the far right gets so close to the axis that the axis and curve share the same screen pixel points. Trace over to see that the function values are getting closer and closer to zero.

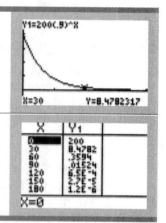

Set △Tbl=30 and look at the table. (The E notation means 6.5E−4 = $6.5*10^{-4}$ = 0.00065.) Thus, the function values are getting closer and closer to zero.

Solving Exponential Equations Graphically

In Section 2.4 we used **Solver** to give us a numerical solution; now we take a graphical approach.

[121]   Suppose that a 200 μg sample of carbon-14 decays according to the formula,

$$Q = 200(0.886)^t,$$

where t is in thousands of years. Estimate when there will be 25 μg of carbon-14 left.

Graph $Y_1=200(.886)^X$ and $Y_2=25$. Set the viewing **WINDOW** to $0 \le x \le 30$ (Xscl=5) and $-10 \le y \le 50$ (Yscl=10). Using **TRACE**, you can arrow to the intersection and estimate $x \approx 17.2$.

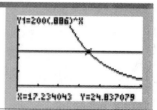

Alternatively, you can use the 2ⁿᵈCALC intersect feature to get a more accurate estimate as shown here.

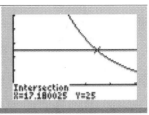

Fitting Exponential Functions to Data

[121]   The population of the Houston Metro area has been growing exponentially. If t is in years since 1900, the table below shows the population in thousands.

t, years since 1900	0	10	20	30	40	50	60	70	80	90	100
P, thousands	184	236	332	528	737	1070	1583	2183	3122	3733	4672

The TI can derive an exponential regression formula in a simple sequence of steps.

Exponential Regression

Enter the data using STAT Edit.

Use 2ⁿᵈSTAT PLOT to turn on the graph and select the lists.

Use ZoomStat to see a data plot. It is clear that a linear regression equation does not seem appropriate.

On a fresh home screen, use STAT CALC and select ExpReg (Exponential Regression) to begin the command. This process is the same as the one we used for linear regression in Section 1.6.

After pasting ExpReg, specify the lists to use for the regression and a Y= equation for the formula to be pasted into.

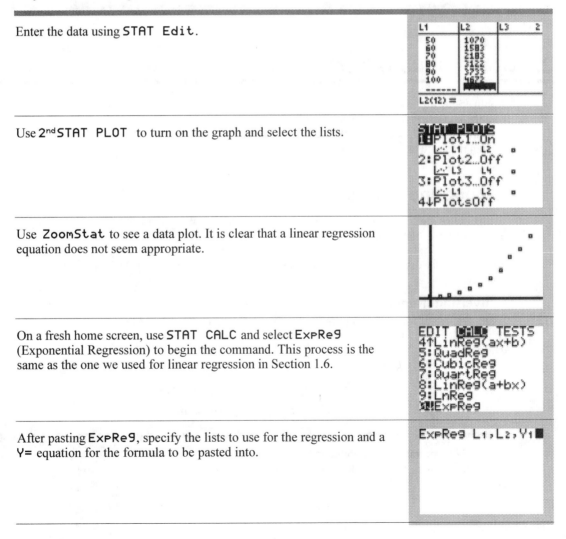

The parameters *a* and *b* are calculated for the equation y=a*b^x. The *r* and r^2 are measures of correlation as previously described.

```
ExpReg
 y=a*b^x
 a=183.5196471
 b=1.034719174
 r²=.9933926192
 r=.9966908343
■
```

Optional step: Here we check that Y₁ has been properly pasted and that both Y₁ and Plot1 have been selected to be graphed.

```
Plot1 Plot2 Plot3
\Y1■183.51964711
95*1.03471917377
84^X■
\Y2=
\Y3=
\Y4=
\Y5=
```

This graph shows the data points and the exponential regression curve.

3.4 Continuous Growth

Applications involving exponential equations are commonly written with base *e* formulas. We need to be able to convert an exponential equation to an equivalent form in another base. So why is this *e* base so important? It models growth that has a continuous growth rate. This is in contrast to a periodic growth rate such as an annual growth rate.

Compound Interest

Growth rates are commonly used in financial applications. We often need to compare rates that are in different units. Annual interest rates are often advertised as percentages in clean decimals; these are called nominal rates. When a nominal rate is compounded more frequently than once a year, then an effective rate is determined. Effective rates often have long strings of decimal digits and need to be rounded.

Converting Nominal and Effective Rates

[127]   What are the effective annual rates of an account paying 12% interest, compounded annually?, compounded monthly? What is the effective rate of an account paying 6% annual interest compounded daily?

    The TI has two conversion commands built in:

$$\blacktriangleright \text{Eff} \langle \textit{nominal rate, compounding periods} \rangle$$

$$\blacktriangleright \text{Nom} \langle \textit{effective rate, compounding periods} \rangle .$$

The ▶Eff⟨ command is available in the catalog. Pressing CATALOG E moves to the items with first letter e. Likewise, CATALOG N moves close to ▶Nom⟨.

On a fresh home screen, calculate the three desired effective rates.

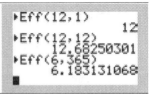

Tip: The financial functions of the TI-83 are much more extensive than is shown here, and they will be quite useful in a business finance course. The interested reader should see Chapter 14 in the *TI–Guidebook*.

---

[128]   The underlying formula relating nominal and effective rates is given as

$$B = P \cdot \left(1 + \frac{r}{n}\right)^{nt},$$

where r is the annual nominal rate and n is the number of times compounding is done in a year. Time, t, is in years and P is the initial deposit.

Letting P be $1 and letting t be 1 year, for a given r and n we can find the annual growth factor. This growth factor equals 1 + effective rate, so that we can easily read off the effective rate from the growth factor.

In the first calculation we find the growth factor of compounding daily with a nominal rate of 6%. Subtract 1 from the growth factor to see that this is the same effective rate shown at the bottom of the previous screen. The second calculation can be read to say that the effective rate of hourly compounding of a 6% nominal rate is 6.1836329%.

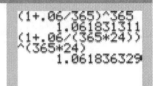

Continuous Compounding Interest

By compounding as ridiculously often as hourly (or every second), the effective rate reaches a limit called the *continuous* rate. It is given as

$$B = Pe^{rt}$$

where r is the annual nominal rate and time, t, is in years, and P is the initial deposit.

Once again for a given r, if we let P be $1 and t be 1 year, we can find the annual growth factor. This growth factor is 1+effective rate, so we can easily read off the effective rate for a continuously compounded rate. This means that we can compare rates of all types.

[129]   Which is better: An account that pays 8% annual interest compounded quarterly or an account that pays 7.95% annual interest compounded continuously?

We compare the growth factors to see that the lower nominal rate compounded continuously is slightly better. The difference would only become significant with large deposits or deposits held for many years.

Tip: Watch out for the word *continuous* in problems with rates. This is the tip-off that you should be using the formula in base e.

Notes:

CHAPTER FOUR

LOGARITHMIC FUNCTIONS

4.1 Logarithms and Properties

Logarithms Are Exponents

[144] We start by confirming that logs are exponents. For example, $\log 100 = 2$ because 2 is the exponent of 10 that gives 100, or $10^2 = 100$.

Here we confirm $10^{\wedge}\log(x)=x$ for $x = 100$. Note the use of **Ans**. In this case it might have been faster to type **10^2**, but in most cases the log value will be a decimal and cumbersome to reenter.

```
log(100)
              2
10^Ans
            100
```

Now a little TI magic trick. Functions such as the log can have a list as input. We first use **{100,.01,30}** as input; the output is a list of function values. Now enter **10^Ans**. Voila! the original list returns. This is just the previous screen done with lists.

```
log({100,.01,30}
)
{2 -2 1.4771212...
10^Ans
        {100 .01 30}
■
```

Solving with Logs

If we want to find a value of Q in $Q = ab^t$ for a given value of t, we say we are *evaluating* Q at t. But when we want a value of t for a given value of Q, we say we are *solving* for t. To solve for t requires that we undo an exponent. This is what the $\log(x)$ and $\ln(x)$ functions do. It is advisable that you be able to solve both exponential and logarithmic functions by using algebra. Solving algebraically requires some decision about whether to use log() or ln() in finding the exact form answer. To find numerical approximations, **solve** and **Solver** require no decision about the kind of log function to use.

Solving an exponential equation where we normally use log

[146] Solve $100 \cdot 2^t = 337,000,000$ for t. To solve algebraically we would normally use the log function, but by using **Solver** there is no need to worry about a method. However we must be satisfied with an approximation.

Enter the equation `0=100*2^X-337000000` in `Solver` and press `ALPHA SOLVE` with the cursor on the `X`.

Solving an exponential equation where we normally use ln

[147]   Solve $5e^{2x} = 50$ for x. Here the ln() function would normally be used to solve algebraically, but the `Solver` command approximates the solution in exactly the same way as the previous example.

Arrow up to the old Solver equation and enter the new equation `0=5e^(2X)-50`. Press `ALPHA SOLVE` with the cursor on the `X`.

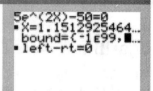

The TI-83 is log friendly

[148]   Find the value of $\log \dfrac{17}{3}$. On some calculators pressing the following five key sequence: log 1 7 / 3 results in the value 0.410, which is incorrect. The TI-83 is foolproof in this regard because it automatically enters a left-parenthesis after log.

First press the five key sequence: `log 1 7 ÷ 3` and `ENTER`. The TI will automatically insert a parenthesis after the log. It also allows you to not close the parenthesis before pressing `ENTER`. The last calculation shows how some calculators would interpret the five key sequence and arrive at 0.410.

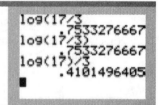

4.2 Logarithms and Exponential Models

There are many applications of logarithms, but their main use is to solve exponential equations. In particular, we show how to find doubling time and half-life for exponential models. We conclude with the important note that logarithms cannot solve all types of exponential equations.

Models

[152]   The US population, P, in millions, is currently growing according to the formula:

$$P = 263e^{0.009t},$$

where t is in years since 1995. When is the population predicted to be 300 million?

| Press MATH 0:Solver, arrow up, and enter the equation shown after the eqn:0= prompt. Notice that the equation is entered as a difference that is set to zero. | ```
EQUATION SOLVER
eqn:0=P-263e^(.0
09T)█
``` |
| --- | --- |
| After pressing ENTER we set the P to 300 and arrow down so that the cursor is on the T variable line. To solve for T, press ALPHA SOLVE. The square dot to the left of T indicates that it has been calculated. | ```
P-263e^(.009T)=0
 P=300
▪T=14.625382497…
 bound={-1ᴇ99,1…
▪left-rt=0
``` |

Doubling Time

[153]   Find the time needed for a turtle population described by the function $P = 175(1.145)^t$ to double its initial size.

For 175 (the initial value) to double, we are solving $2 \cdot 175 = 175(1.145)^t$ for t. However, after a division by the initial value, this is the same as solving $2 = (1.145)^t$. Similarly for any exponential model, $Q = ab^t$, the doubling time is found by solving $2 = b^t$. The solution is $t = \dfrac{\log(2)}{\log(b)}$ and can be found with your calculator.

| On this screen we see that the solution to the turtle doubling time is independent of which logarithmic function we use. Our turtle population doubles roughly every five years. | ```
log(2)/log(1.145
)
       5.119080084
ln(2)/ln(1.145)
       5.119080084
``` |
| --- | --- |

Half-life

[154]   While doubling time is found for an increasing exponential function, half-life is commonly found for an exponentially decaying function. For example, carbon-14 decays at an annual rate of 0.0121%. Find the half-life of carbon-14.

| In general, half-life is found by solving $\frac{1}{2} = b^t$ for $t$. We note that the growth rate is $b = 1 - 0.000121$, and we solve using the log (or ln) function. The half-life of carbon-14 is 5728 years. (See the tip below if you are confused by 0.0121% = 0.000121.) | ```
log(1/2)/log(1-.
000121)
       5728.142515
``` |
| --- | --- |

Tip: Use care when entering percentages; they must be entered as their decimal equivalent. For example, 5% = 0.05, and for a fractional rate, 0.5% = 0.005.

Tip: It is safer to enter expressions without simplification. They are easier to double check later for correctness.

Applications involving exponential equations are commonly written with base e formulas. We need to be able to convert an exponential equation to an equivalent form in another base. So why is this e base so important? It models growth that has a continuous growth rate. This is in contrast to a periodic growth rate such as an annual growth rate.

Converting bases

[155]   Convert the function $P = 175(1.145)^t$ to the form $P = ae^{kt}$. Convert the function $P = 7e^{0.3t}$ to the form $P = ab^t$.

In general, the equality of the two forms means $ab^t = ae^{kt}$. Since a is the same in both cases, the equality becomes $b^t = e^{kt}$. With t also being the same, we further simplify to see that we need only solve $b = e^k$. In the first example the solution of $1.145 = e^k$ is $k = \ln(1.145)$. In the second we have $b = e^{0.3}$. Seeing that the conversion is finding k from b or b from k means a simple calculation on the home screen. However, you might want to show off and use `Solver` as a single technique that converts either way.

Enter the general equation as a difference.	`EQUATION SOLVER` `eqn:0=B-e^(K)`

Here we set **B** and solve for **K**. This gives the same answer as entering `ln(1.145)` on the home screen.	`B-e^(K)=0`
	`  B=1.145`
	`■ K=.13540463700…`
	`  bound={-1ᴇ99,1…`
	`■ left-rt=0`

To solve the second equation, we set **K=.3** and solve for **B**. This gives the same answer as entering `e^(.3)` on the home screen.	`B-e^(K)=0`
	`■ B=1.3498588075…`
	`  K=.3`
	`  bound={-1ᴇ99,1…`
	`■ left-rt=0`

Not All Exponential Equations Can Be Solved by Logarithms

[156]   With t in years, the population (in millions) of a country is given by $P = 2(1.02)^t$, while food supply (in millions of people who can be fed) is given by $N = 4 + 0.5t$. Determine the year in which the country first experiences food shortages.

If we try to algebraically solve $4 + 0.5t = 2(1.02)^t$ using logs, we cannot isolate t; however, there is a numerical solution. (In fact, there are two such solutions.) If we blindly use `solve`, we may find an unacceptable solution; the relevant solution is the 199th year.

Use `solve` with the difference of the two formulas. In the first try, the guess for **X** is **0**. This gives a negative answer. We change the guess and find the other solution. (Use **2ⁿᵈENTRY** to avoid tedious reentry.)	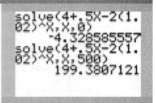

Tip: In the previous example, it would have been safer to use a graphical approach, since you
would then see where the graphs intersect and be assured that you have a relevant solution.
Safety has its price: A graphical approach has more steps.

The difference between log and ln

Why are there two logarithm keys? The answer is that `log` undoes exponents to the power of 10,
while the other `ln` undoes exponents to the power of e. Either one can be used in solving any
exponential equation. However, the ln is simpler for exponential equations involving e, and the log is
preferred for equations with powers of 10.

Consider solving the previous equation $300 = 263e^{0.009t}$. You can verify
using algebra and logarithm properties that a solution using ln is: $t =$
$\ln(300/263)/0.009$. If you used log to undo the exponent, then the
solution is written $t = \log(300/263)/(0.009\log(e))$. The screen shows that
these are the same solution.

Tip: The value of e can be pasted from the keypad with 2nd e (above divide) or pasted from the
CATALOG.

4.3 The Logarithmic Function

The Graph of ln(x)

This screen shows the graph of `Y1=ln(X)` using the `Zdecimal`
window. When **TRACE** is used we see that for `X=0` there is no `Y` value.
This means that the function is undefined at this point. By tracing to the
left, you will see that it is also undefined for negative values of x.

Trace to the right and `Y` values are shown. In this case we see that the ln
function has a zero (also called a root) at $x = 1$. The property is listed as
$\ln 1 = 0$.

How can we show the property $\ln(e) = 1$? If you trace by arrow keys and
attempt to find `X` when `Y=1`, you must settle for an approximate result
such as that shown on the bottom line. It says: $\ln(\approx e) = \approx 1$. In trace
mode, you can set `X=e` and see that $\ln(e) = 1$.

The Graph of ln(x) and e^x

[137]   The two functions $y = \ln(x)$ and $y = e^x$ are inverses. Their graphs are reflections of one another across the line $y = x$.

Enter three functions: Y₁=ln(X), Y₂=e^(X), and Y₃=X. Graph with a ZDecimal window.

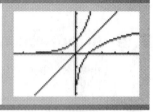

> *Tip:*   ZDecimal was used in the previous example because it shows the $y = x$ line at a true 45° angle. If you use a ZStandard window, the graphs will not look like perfect reflections. But you can use ZOOM 5:ZSquare to square up any given window.

Table Values for ln(x)

What can we expect to see from a table of ln values?

Here we proceed showing both Y₁ and Y₂ from the previous example. We know the ln function is undefined for $x = 0$, and this is denoted as ERROR. For this table the ⊿Tbl has been set to 1. We note that both of these functions are increasing, with the ln function growing rather slowly.

X	Y₁	Y₂
0	ERROR	1
1	0	2.7183
2	.69315	7.3891
3	1.0986	20.086
4	1.3863	54.598
5	1.6094	148.41
6	1.7918	403.43

Y₁▪ln(X)

To find self-selected values for a table, we use TBLSET and set Indpnt: Ask. This means that the independent variable x will be asked for. (It will ignore the settings for TblStart and ⊿Tbl.)

TABLE SETUP
TblStart=0
⊿Tbl=1
Indpnt: Auto **Ask**
Depend: **Auto** Ask

Here we see a screen showing that we have manually requested the values for 0, 1, and 2. The bottom of the screen shows that we have pressed 2ⁿᵈe, and after pressing ENTER we will find out $\ln(e) = 1$.

X	Y₁	Y₂
0	ERROR	1
1	0	2.7183
2	.69315	7.3891

X=e

> *Tip:*   With Indpnt: set to Ask, you can mix the order or size of x table entries. Even special values like π and e can be used.

> *Tip:*   If in the future you are surprised to see a blank table after pressing TABLE, then you probably forgot to turn Indpnt: back to Auto in TBLSET.

4.4 Logarithmic Scales

Log-Log Scales

[170]   Plot the table below, which shows the average metabolic rate in kilocalories per day for animals of different weights.

Animal	Weight(lbs)	Rate(kcal/day)
Rat	1	35
Cat	8	166
Human	150	2000
Horse	1750	9470

Use **STAT Edit** to enter the data into two lists as shown.

Use **2ⁿᵈ STAT PLOT** to set **Plot1** as shown. (Don't forget to be sure that the **Y=** defined functions are turned off.)

Use **ZoomStat** to set a good window for the data.

This does not look like a very good plot because some of the values are squeezed at the origin.

The remedy is to change to a Log-Log scale. This means that we will graph the data, with both *x*- and *y*-axes being in log scale. We convert **L1** to log values and store them in **L3**. (Arrow up to highlight **L3**, and then enter the command shown on the bottom definition line.)

Also convert L2 to log values and store them in L4.

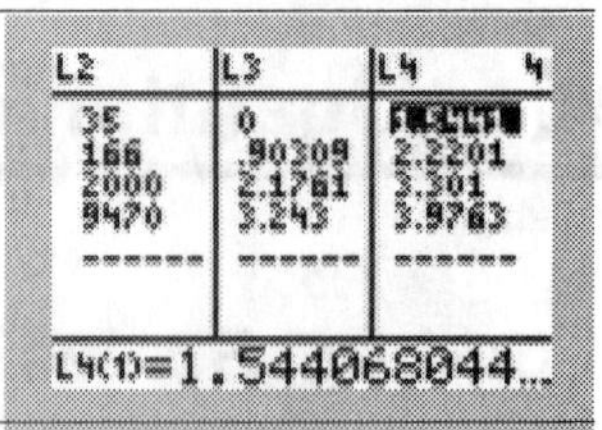

L2	L3	L4	4
35	0	1.5441	
166	.90309	2.2201	
2000	2.1761	3.301	
9470	3.243	3.9763	
-----	-----	-----	

L4(1)=1.544068044...

Turn off Plot1, and set up Plot2 as shown here. Press ZoomStat. A new window is needed because the data are now converted to log form.

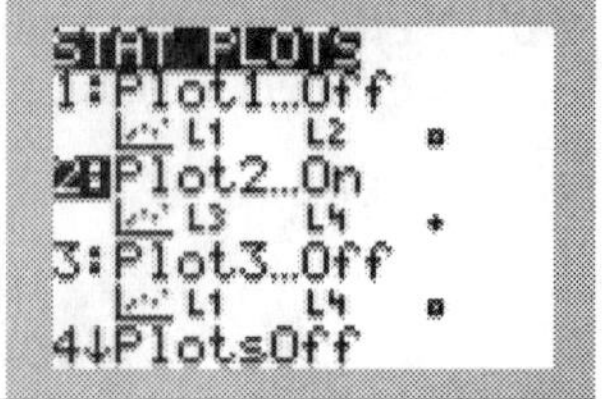

STAT PLOTS
1:Plot1...Off
  L1 L2
2:Plot2...On
  L3 L4
3:Plot3...Off
  L1 L4
4↓PlotsOff

This is quite nice to see the points spread out. They also appear to be approximately linear. This is a tip-off that the original (non-log) relationship could be modeled by a power function.

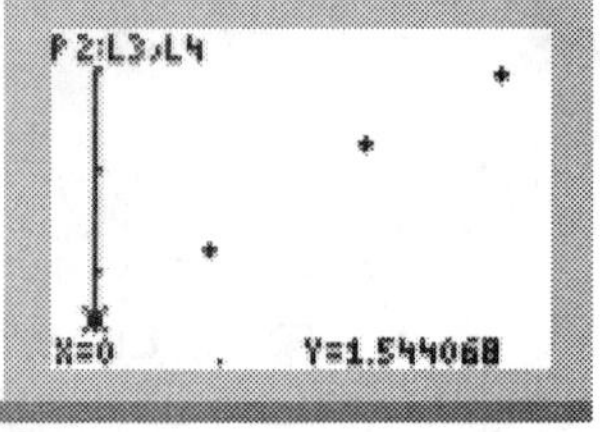

P2:L3,L4

X=0   Y=1.544068

Tip: In the last screen TRACE was used to identify the plot and its data points. When both a Plot and Y= function are shown on the same screen, the up/down arrow will allow you to switch between a curve and a plot of points.

In Section 1.6 we saw how to fit a linear regression equation to data. In this section we will find out how to fit either an exponential regression equation to data.

The Sale of Compact Discs

[166]   The table below shows the fall in sales of vinyl long playing records (LPs) and the rise of compact discs (CDs) during the years 1982 through 1993. We use $t = 1$ for 1982.

t, years since 1982	c, CDs (millions)	l, LPs (millions)
0	0	244
1	0.8	210
2	5.8	205
3	23	167
4	53	125
5	102	107
6	150	72
7	207	35
8	287	12
9	333	4.8
10	408	2.3
11	495	1.2

The textbook's approach to fitting a curve to this data is to use the log function to *linearize* the data. This yields a linear regression equation for the converted data that can then be used to algebraically derive the exponential regression formula. This approach is good for conceptual understanding, and your instructor may want you to do the work that way. However, the TI can derive the exponential regression formula in a simple sequence of steps.

Exponential regression: linearized by a log scale

Enter the data using STAT Edit. Arrow to the header of L3. Press ENTER to be prompted for the formula shown at the bottom of the screen.

L1	L2	▦	3
0	244	------	
8	210		
5.8	205		
23	167		
53	125		
102	107		
150	72		

L3 =ln(L2)■

Now L3 has the ln values of each entry in the L2 list.

L1	L2	L3	3
0	244	5.4972	
8	210	5.3471	
5.8	205	5.323	
23	167	5.118	
53	125	4.8283	
102	107	4.6728	
150	72	4.2767	

L3(1)=5.497168225...

On a fresh home screen, use STAT CALC and select LinReg(a+bx) (Linear Regression) to begin the command. This process is the same as the one we used in Section 1.6.

```
EDIT CALC TESTS
2↑2-Var Stats
3:Med-Med
4:LinReg(ax+b)
5:QuadReg
6:CubicReg
7:QuartReg
8↓LinReg(a+bx)
```

After pasting LinReg(a+bx), specify the regression lists as the first and the third, L1 and L3. We are finding a linear regression on L1 and L3=ln(L2).

```
LinReg(a+bx) L1,
L3■
```

The parameters a and b are calculated for the equation y=a+bx. But this is really $\ln(y) = a + bx$ so that to reach an exponential regression equation we take both sides to the power of e.

```
LinReg
y=a+bx
a=5.518857815
b=-.0109169351
r²=.9858422773
r=-.9928959046
■
```

We calculate $e^a \approx 250$ and round b to define Y1. With both Y1 and Plot1 selected we graph in a ZoomStat window.

```
e^(5.52)
         249.6350372
■
```

The graph shows the data points and the exponential regression curve.

Exponential regression: Using built-in regression equation

| Using only L₁ and L₂ of the previous example, we now apply the TI's built-in ExpReg to reach the same result. On a fresh home screen, use STAT CALC and select ExpReg | ```
EDIT CALC TESTS
6↑CubicReg
7:QuartReg
8:LinReg(a+bx)
9:LnReg
0:ExpReg
A:PwrReg
B↓Logistic
``` |
| This specifies calculating an exponential regression using L₁ and L₂ and pasting the formula into Y₁. | ```
ExpReg L₁,L₂,Y₁■
``` |
| We see the coefficients, but it is still unclear whether we have the same formula since the previous one was given in the form $y = ae^{kx}$ and this one is of the form $y = ab^x$. | ```
ExpReg
 y=a*b^x
 a=249.3500705
 b=.9891424384
 r²=.9858422773
 r=-.9928959046
■
``` |
| We see that the two forms of the equation are essentially the same. | ```
e^(-.011)
       .9890602788
■
``` |
| Since Y₁ has been pasted as part of the ExpReg command, we can press GRAPH and see the same plot and curve as derived by the previous method. | |

CHAPTER FIVE

TRANSFORMATIONS OF FUNCTIONS

5.1 Vertical and Horizontal Shifts

The graph of a function may be shifted vertically or horizontally. In this section we look at the change in a function's formula that produces a shift of its graph.

Inside and Outside Changes

[185]   If $n = f(A) = \dfrac{A}{250}$ gives the number of gallons of paint needed to cover a house of area A ft^2, show the difference between $f(A+10)$ and $f(A)+10$ numerically.

Enter Y₁ in the Y= menu. Now define Y₂ and Y₃ in terms of Y₁ as shown. Recall that the Y₁ used in the Y₂ definition must be pasted from the VARS Y-VARS 1:Function screen.

Before pressing 2ⁿᵈTABLE, use 2ⁿᵈTBLSET to start the table at $x = 0$, and have it go up by 10 for each x. In the table shown, the cursor has been moved to exhibit the Y₂ formula. Looking at the table, we see that a value of Y₁ appears back 10 units (one line up) in the Y₂ table.

Return to the Y= screen and turn off Y₂. Now press 2ⁿᵈTABLE to compare Y₁ and Y₃ values. Here we see that each Y₃ value is just the Y₁ values plus 10.

Tip: To evaluate a function for a single value, the Y₁(0) method is quite handy, but if you wanted to see many values, say from 0 to 12, then a table is the best choice.

Horizontal Shifts

Graph $f(x) = x^2$. Graph f shifted to the right two units.

Define the function f as Y_1. Shifting to the right means that the variable x is replaced by $x–2$. So the shifted graph can be defined as $Y_2=Y_1(X-2)$.

```
Plot1  Plot2  Plot3
\Y1■X^2
\Y2■Y1(X-2)
\Y3=
\Y4=
\Y5=
\Y6=
\Y7=
```

Use a `ZDecimal` window to see the two graphs. Indeed the graph of Y_2 is the graph of Y_1 shifted two units to the right. Although not used here, the TI-83 graph style feature would allow you to draw Y_2 as a thicker graph so that you can visually distinguish between the two graphs.

Vertical Shifts

Graph $f(x) = x^2$. Graph f shifted down one unit.

With the function f as Y_1, shifting down one unit means the value of the function Y_1 must be one unit less. So the shifted graph can be defined as $Y_3=Y_1(X)-1$. In this example we turned off the Y_2 definition.

```
Plot1  Plot2  Plot3
\Y1■X^2
\Y2=Y1(X-2)
\Y3■Y1(X)-1
\Y4=
\Y5=
\Y6=
\Y7=
```

Using a `ZDecimal` window we see the two graphs. Indeed, Y_3 is just Y_1 shifted one unit down.

Combining Shifts

[187]   With $f(x) = x^2$, graph f shifted to the right two units and down one unit.

Here we make a subtle change to the definition of Y_3; it is defined in terms of Y_2. Thus Y_3 depends on Y_2, and Y_2 depends on Y_1. Note that even though Y_2 is not selected to be graphed, it can still be used as part of another definition.

```
Plot1  Plot2  Plot3
\Y1■X^2
\Y2=Y1(X-2)
\Y3■Y2(X)-1
\Y4=
\Y5=
\Y6=
\Y7=
```

Using the `ZDecimal` window we see the two graphs. We see Y_3 as Y_1 shifted two units to the right and one unit down.

Checking formulas

Find a formula for the combined shifts shown above and graph both functions to check that they are the same.

Replacing the variable x by $x-2$ and then subtracting 1, we have the combined shift formula shown as Y_4. Since we want to check that the two graphs Y_3 and Y_4 are the same, we need some way of knowing that the second function is graphing the same curve as the first. After all, it could just be graphing outside the window. We use the graph style, ⌀, which draws a circular cursor as it graphs.

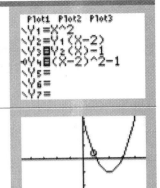

After the first graph is drawn, the second graph can be seen as the same because the circular cursor traces out the same points. (See the tip below if you want to make the cursor pause as in this example.)

Tip: Press **ENTER** to pause a graph and then **ENTER** again to resume graphing. The pause indicator is twinkling dots in the upper-right corner of the screen. Press **ON** to break and abort active graphing or calculating.

5.2 Reflections and Symmetry

A reflection over the x-axis corresponds to an outside change, while a reflection over the y-axis corresponds to an inside change. Symmetry occurs when a reflection gives the same graph.

Reflections

[191] The table below gives values of a function f. Write a formula for $f(x)$ and graph f.

x	-3	-2	-1	0	1	2	3
$f(x)$	1	2	4	8	16	32	64

In this example Δx is constant, and it is easy to see that the values of $f(x)$ are doubling. We identify $f(x)$ as an exponential function with growth factor 2. Also we see that the initial value (when $x = 0$) is 8. Thus our formula is $f(x) = 8(2)^x$.

Set the viewing window to $-3 \le x \le 3$ (Xscl=3) and $-65 \le y \le 65$ (Yscl=10) and graph f. It shows the graph is increasing and concave up.

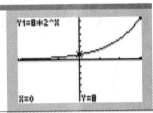

To reflect the graph f across the x-axis, graph $g(x) = -f(x)$.

In this and the next two screens, we use the graph style feature to make the reflected graph show as a thick curve. Recall that this style is set in the Y= screen on the far left of the definitions.

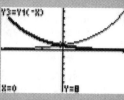

To reflect the graph f across the y-axis, graph $h(x) = f(-x)$.

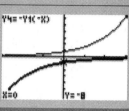

To reflect the graph f across the origin, graph $k(x) = -f(-x)$.

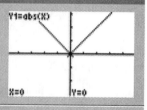

Symmetry

[195]   Determine whether the following functions are symmetric across the y-axis, the origin, or neither.

(a) $f(x) = |x|$ (b) $g(x) = 1/x$ (c) $h(x) = -x^3 - 3x^2 + 2$

Enter these functions as Y1, Y2, and Y3. Set the viewing window to ZDecimal for the graphing. We see the graph of $f(x)$ suggests the function is symmetric about the y-axis. Algebraically, we can verify that $f(-x) = f(x)$. You may also want to check by graphing Y4=Y1(-X) and comparing it to the graph of Y1.

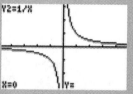

The graph of $g(x)$ suggests that the function is symmetric about the origin. Algebraically we can verify that $-g(-x) = g(x)$. You may also want to check by graphing Y4=-Y2(-X).

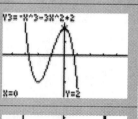

The graph of $h(x)$ suggests that the function has no symmetry. Algebraically, we can verify that $h(-x) \neq h(x)$ and $-h(-x) \neq h(x)$.

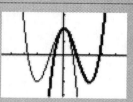

We see that $h(-x) \neq h(x)$ by graphing Y4=Y3(-X) with Y4 in thick graph style. The two graphs are clearly not the same.

5.3 Vertical Stretches and Compressions

A vertical stretch or compression is represented by an outside change to its formula.

Stretch Factor

[200] A yam is placed in a 300°F oven. The table gives $r(t)$, the yam's temperature t minutes after being placed in the oven. Make a table of values for $q(t) = 1.5r(t)$. Graph the functions.

t	0	10	20	30	40	50	60
$r(t)$	0	150	225	263	281	291	295

Enter the data in L1 and L2; then create a new list L3=1.5*L2. To calculate this L3, arrow up to highlight the L3 header and press ENTER. The cursor appears on the bottom line to enter the calculation. Press ENTER again and the calculated numbers appear in L3.

Since we have lists and but no formulas, we must use STAT PLOT to graph these values. Press 2ⁿᵈ STAT PLOT to see the current settings for the three possible plots. Select the settings shown. (The intermediate plot setting screens are not shown here.) An xyLine plot has been selected; it is similar to a scatter plot, but connects consecutive data points with a line.

After the settings are made, use ZoomStat to graph the plots. We see that the second graph is a stretch of the first graph. (If this graph has extra curves, you may have forgotten to turn off function graphing in the Y= list.)

Average Rate of Change

[201] How is the average rate of change of a function affected by a stretch? We show numerically that if $g(x) = k \cdot f(x)$ then on any interval,

$$\text{Average rate of change of } g = k \cdot (\text{Average rate of change of } f).$$

We start with the three lists from the previous example. In L4 we will calculate the average rate of change for the original function. Since $\Delta x = 10$ for all the intervals, we divide by 10 and avoid using ∆List(L1).

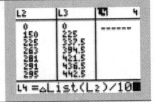

In L5 we will calculate the average rate of change for the stretched function.

If you are arithmetically swift, you will see that L5=1.5*L4. To check this, you can use L6=1.5*L4. You will find that L5 is the same as L6.

5.4 Horizontal Stretches and Compressions

[210]   Match functions $f(t) = e^t, g(t) = e^{0.5t}, h(t) = e^{0.8t}, j(t) = e^{2t}$ with the graphs.

Enter the four functions in the Y= menu.

Set the viewing window to $0 \le x \le 10$ (Xscl=1) and $-20 \le y \le 100$ (Yscl=50) and graph. Without using trace, see if you can identify which graph is which. Here TRACE has been turned on, and the cursor has been moved to identify the first curve.

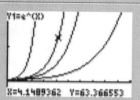

The order of the functions from left to right is Y4, Y1, Y3, Y2. The Y4 has been squeezed toward the y-axis in comparison to Y1, since it is happening twice as fast. The function Y2 is stretched twice as far from Y1 since it is happening half as fast.

The symmetry of stretching

In the previous example, you may feel that the whole function has just been shifted left since the values of the negative domain are hidden as they hug the x-axis. This is not the case, a stretch will occur away from the y-axis on both sides. To see this we graph the absolute value function and two related functions.

$$\text{(a) } f(x) = |x| \quad \text{(b) } g(x) = |2x| \quad \text{(c) } h(x) = |.5x|$$

Enter these functions as Y1, Y2, and Y3. Set the viewing window to ZDecimal. We see that the graph of g is compressed on both sides, and the graph of h is stretched on both sides.

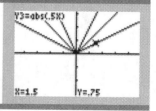

5.5 Quadratics

The standard form of a quadratic equation is $y = ax^2 + bx + c$, where a, b, c are constants, $a \neq 0$. Letting $a = 1$ and $b = c = 0$, we have the simple quadratic equation $y = x^2$. It can be shown that any quadratic equation is just a transformation of this simple equation. The basic shape is called a parabola.

Transformation of the Basic Quadratic

[214]   Let $f(x) = x^2$ and $g(x) = -2(x+1)^2 + 3$. Write g as a function of f. Graph f and g.

It is easy to see that $g(x) = -2f(x+1) + 3$. Enter that definition as Y_2 and press ZDecimal.

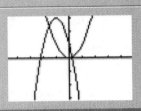

We see that g is a transformation of f. You may also want to graph $Y_3 = -2X^2 - 4X + 1$ to see that Y_3 is the standard form of Y_2.

Applications of Quadratics

Maximum/minimum

[218]   For t in seconds, the height of a baseball in feet is given by the formula

$$y = f(t) = -16t^2 + 64t + 3.$$

Find the maximum height reached by the baseball.

Enter the formula in Y_1 and graph with a viewing window of $0 \leq x \leq 4.7$ (Xscl=1) and $-20 \leq y \leq 100$ (Yscl=10). Use TRACE to find the maximum value as shown. By using the right half interval of the ZDecimal, the trace feature will show nice values of x. Good settings for the y values are found by trial-and-error or by using ZoomFit.

In some cases, the trace values may not land on the x value that gives the desired accuracy. In such cases, you can use maximum from the 2nd CALC menu. This process requires three prompts before it finds the maximum. You must specify:

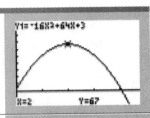

Left Bound — a point to the left of the maximum

Right Bound — a point to the right of the maximum

Guess — a point near the maximum.

We see here the same answer. The maximum height is 67 feet. This maximum height is at the vertex of the parabola.

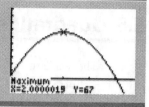

> *Tip:*   Finding a vertex (maximum/minimum) from a quadratic graph may show numbers that need some common-sense rounding. For example, it may find Y=2.99999999 when the exact answer is 3. In the above example, X=2.0000019 should be X=2.

The zeros of a quadratic

[218]   A second question from the baseball model is when does the ball hit the ground. This is found when $f(t) = -16t^2 + 64t + 3 = 0$, that is, when $f(t)$ has a zero.

Using the previous settings, we can trace and get a close approximation.

Get a more accurate solution from **zero** on the 2nd **CALC** menu. Like the maximum process you are required to answer three prompts (**Left Bound, Right Bound, Guess**) before it finds the zero.

Notice that the y value is not exactly zero. But,

$y = -1E-11 = -0.00000000001$ is pretty close.

The zeros of a quadratic function can be found by using our program **QF** from Section 2.6.

CHAPTER SIX

TRIGONOMETRIC FUNCTIONS

6.1 Introduction to Periodic Functions

Definition: A function is *periodic* if its values repeat at regular intervals. In functional notation this means that there is a c such that $f(t+c) = f(t)$ for all t in the domain of f. We see that the graph of the function repeats itself over successive intervals of length c. We start with an example of such a graph.

An Example of a Periodic Function

Show that the function $f(x) = x - \text{int}(x)$ is periodic. The function, $\text{int}(x)$, is the integer part of x. For example, $\text{int}(2.7128) = 2$.

When entering the function definition for Y₁, the int(x) function needs to be pasted from the MATH NUM menu that is shown here or it can be found as int in the CATALOG.

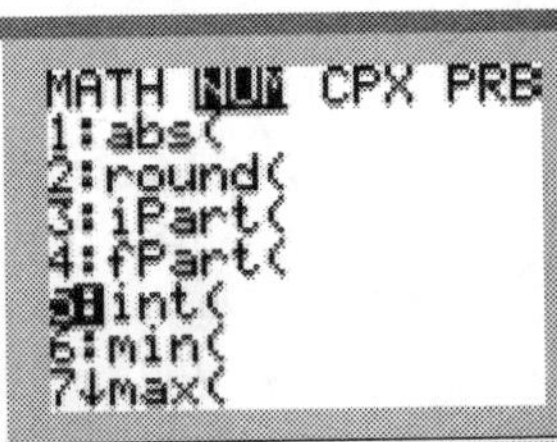

After entering the function, graph in a ZDecimal viewing window. For the screen shown here, the MODE was set to Dot instead of Connected and then TRACE was turned on to display the function definition. The periodic behavior is evident.

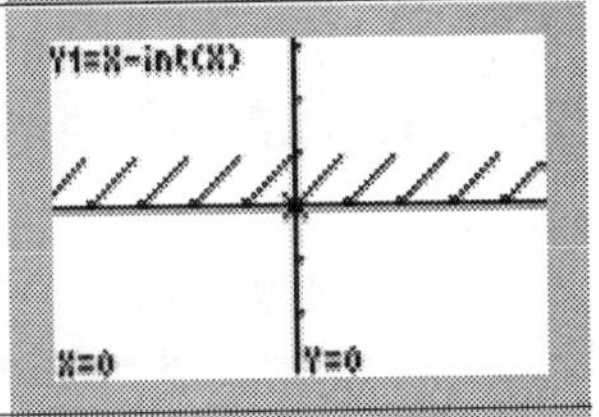

Take special care with table values of periodic functions. Since the period is 1, if we set TblStart=0 and ⌂Tbl=1, then the table looks like the function is constant.

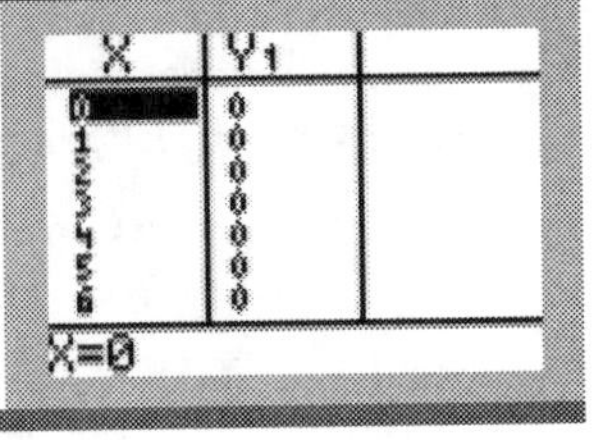

An Example of a Non-periodic Function

Enter `Y₂=abs(int(X))` to see a function whose graph (in a `ZDecimal` window) is made up of repeated horizontal line segments of length 1. So it is clear that the only likely candidate for the period length is $c = 1$. However, it is not periodic since $f(x+1) \neq f(x)$.

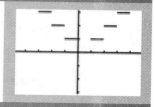

6.2 The Sine and Cosine Functions

On a unit circle, each angle θ specifies a point, $P = (x, y)$. The sine function of the angle θ is the y-coordinate of the point P. The cosine function of the angle θ is the x-coordinate of the point P.

Degree Mode

Use the `MODE` key to change the angle mode to `Degree`.

```
Normal Sci Eng
Float 0123456789
Radian Degree
Func Par Pol Seq
```

Exit and check the mode by finding a known value like sin(90°)=1.

```
sin(90)
             1
```

Tip: On many scientific calculators, the angle mode is shown on a status line across the bottom of the screen. There is no such indicator on the TI-83. You must use `MODE` to check.

The Ferris Wheel

[244] A ferris wheel has a radius of 225 feet. Find your height above the ground as a function of the angle θ, measured from the 3 o'clock position. What is your height when the angle is 0°, 30°, 60°, ..., 360°?

We want the sequence of angle values in `L1`. There is an easy way to enter sequences into a list. From a clean home screen, press `2ⁿᵈ LIST` and view the `OPS` menu as shown here. Select `5:seq(`. The syntax is:

$$\text{seq}(expression, variable, begin, end, increment).$$

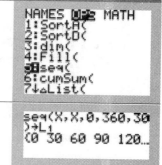

Complete the command as shown to store the list in `L1`. The same list could have been generated in an alternative way by using `seq(30X,X,0,12,1)`.

Next we want height values stored in L2. This screen shows the actual values in L2 which were generated by the formula

Check 2ⁿᵈSTAT PLOT to set Plot1 as shown here. The plot is set On, it draws a scatter plot using L1 and L2, and the data points will be square markers. If these are not the settings, then edit the Plot1 setup.

Use ZoomStat to see a scatter plot. The axes appear as double lines because the Xscl and Yscl are so small that the tick marks form a second solid line on each axis.

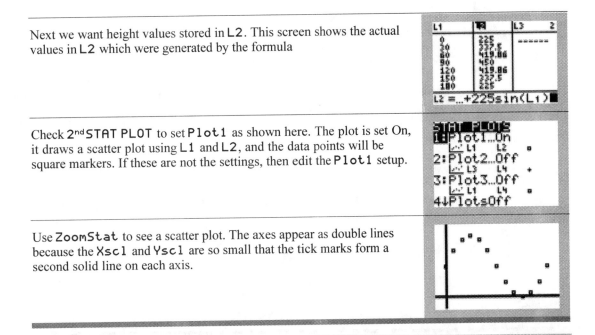

Tip: The seq(command should not be confused with seq as a function mode on the MODE screen.

6.3 Radians

In the previous section we stipulated that all the angles be in degrees. For calculus, radians are commonly used as a unit of angle measure. In this section we see how to convert back and forth between the two measures.

Converting Between Degrees and Radians

[248]   Convert 3 radians to degrees. Convert 3 degrees to radians.

To convert to degrees, check that the mode is set to degree. Return to a fresh line on the home screen and enter 3. (This home screen entry is not shown.)

Press 2ⁿᵈANGLE and select 3: ͬ, which pastes the radian designation to the home screen.

You see that a 3 radian angle is about 172 degrees.

To convert degrees to radians, change to Radian mode. Enter 3 and use 2ⁿᵈANGLE and 1: °, which pastes the degree symbol to the home screen. You see that 3 degrees is about 0.05 radians.

> *Tip:* The previous screen shows two conversions but it does not show that the angle mode had been changed in between the two calculations.

[250]   What is the length of arc cut off by an angle of 120° on a circle of radius 12 cm?

| By using: Arc length = $r\,\theta$, with θ in radians, this type of problem is easy. The conversion to radians is even done automatically by using the degree symbol. The exact value is 8π. Without the calculator, this could be obtained by converting the special angle to radians ($120° = 2\pi/3$) and multiplying by 12. | ```12*120°   25.13274123 8π   25.13274123``` |

> *Tip:* As a rule it is best to leave the angle mode set to radians and write angles that are in degrees with the degree symbol designator (as shown in the previous screen).

6.4 Graphs of the Sine and Cosine

Exact Values for Sine and Cosine

[253]   In the last example of the previous section we saw that $8\pi \approx 25.13274123$. The calculator gave us an approximation instead of the exact form 8π. In exact terms, $\cos(30°) = \sqrt{3}/2$; however, the calculator will show $\cos(30°) = 0.8660254038$. In some cases, your instructor may want you to write an answer in exact form. More advanced calculators, like the TI-89 and TI-92, have the capability to leave answers such as $\cos(30°)$ in exact form.

Graphing a Sine Function

[255]   Graph the ferris wheel function giving your height, $h = f(\theta) = 225 + 225\sin(\theta)$, in feet, above the ground as a function of the angle θ.

Although radians are usually recommended for graphing we do this example in **De9ree** angle mode. This allows us to recycle our scatter plot from Section 6.2.	```Normal Sci Eng Float 0123456789 Radian Degree Func Par Pol Seq```
From the previous ferris wheel encounter in Section 6.2 we have **L1** and **L2** used for a scatter plot with $\theta = 0°$, 30°, 60°, ..., 360°. Leave **Plot1** selected and define **Y1** $= f(\theta)$.	```Plot1 Plot2 Plot3 \Y1☐225+225sin(X)☐ \Y2= \Y3= \Y4= \Y5= \Y6=```
Extend the **WINDOW** settings to those shown to see more of the periodic behavior.	```WINDOW Xmin=-30 Xmax=990 Xscl=90 Ymin=0 Ymax=500 Yscl=50 Xres=1```

Press **GRAPH** and the periodic function as shown. We see a period of 360, a midline of $h = 225$, and an amplitude of 225.

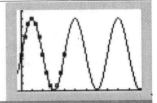

Tip: For graphs of the pure trigonometric functions, such as $y = \sin(x)$ and $y = \cos(x)$, the **ZTrig** graph window is a good first setting. If **MODE** is set to **Degree**, the **X** settings will be in degrees.

6.5 Sinusoidal

Definition: A *sinusoidal* function is of the form

$$y = A\sin(B\,(t - h)) + k \ \text{ and } \ y = A\cos(B\,(t - h)) + k.$$

The general theory of transformations from Chapter 5 outlines the effect of the parameters on the basic sine or cosine function. We use graphing to compare the effect of a change in a parameter.

Horizontal Shifts

[261]   Compare the graph of $f(t) = \cos(3t)$ and $g(t) = \cos(3 - \pi/4)$.

Enter the two functions and graph in a **ZTrig** window. **TRACE** shows that a maximum of the **Y1** graph occurs at $x = 0$.

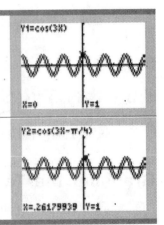

Use the up arrow to change which function is being traced. Arrow right to see that the **Y2** graph has a maximum at $x \approx 0.26$. Thus, the second graph is the same as the first graph but shifts right by ≈ 0.26. Looking at the **Y2** definition, you might expect the shift to be $\pi/4$ (≈ 0.79). This example reminds us to factor when finding the horizontal shift. Here $g(t) = \cos(3(t - \pi/12))$ and $\pi/12 \approx 0.26$.

Tip: The **ZTrig** setting has an advantage in trace because it uses x-values that are fractions of π.

An Application of the Transformed Sine Function

[263]   In previous sections, we dabbled with the ferris wheel height problem by using a rotation angle. In reality, the height of the London ferris wheel is more naturally dependent on time. We now model height by using a sinusoidal function that assumes one revolution takes 30 minutes and that we board at the bottom. You can verify that the appropriate formula is

$$f(t) = 225\sin(\tfrac{\pi}{15}\,(t - 7.5)) + 225.$$

In Y₁ we have been cautious and inserted extra parentheses. Since we know that the midline is $y = 225$, we use Y₂ to graph it. The graph style of Y₂ is dotted (·.).

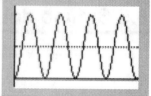

Because it is an application, we have a good idea about the length of time (x-values) to show, say 120 minutes. In this example we also know the y-values because the wheel height is 450 feet. (As usual, we pad the top and bottom of the y settings.) We also set Xres=2. This is a feature that speeds up graphing by only plotting every other x-value.

The Xres (x-resolution) feature was used here not to speed up the graphing but to cause the midline to show as a dotted line. In dot style, a horizontal line looks like a connected solid line. The graph of Y₁ is what is important here. The midline graph is cosmetic. Using trace you can see your height as a function of time.

> *Tip:*   Xres can be used to speed up graphing, but be careful about using a setting that is too high since it may distort a nonlinear graph. (Try Xres=5 in the previous case.) Reset Xres after use. (ZTri9 will not reset XRes.)

> *Tip:*   Trigonometric graphs often show the pattern early in the graphing process. If you realize the graph is not what you want before it finishes, press ON to break.

6.6 Other Trigonometric Functions

In addition to the sine and cosine functions, there are four other common trigonometric functions.

Graphing the Tangent Function

[268]   The tangent function is the quotient of the sine and cosine. This function is used enough that it has its own key, TAN. Graph $y = \tan(\theta)$ in radian mode.

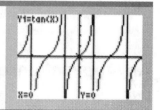

Check the MODE to ensure a Radian setting. Enter the function in Y₁ and graph with ZTri9. This closely matches the textbook graph of the tangent function, but there is an important difference. The four vertical lines at $-3\pi/2$, $-\pi/4$, $\pi/4$, and $3\pi/4$ are dotted to indicate that they are vertical asymptotes. In this TI graph, the four vertical lines are connecting true discontinuities as explained in Section 2.3.

Change **MODE** from **Connected** to **Dot** and graph again. Use the right arrow to trace function values. The next point to the right of the one shown will be $x \approx 1.44$ and $y \approx 7.6$ (it is outside the viewing window.) One more arrow to the right will show $x \approx 1.57$ and $y \approx -4.88$ and it is the 7.6 and –4.88 that were connected as a vertical line in the previous graph.

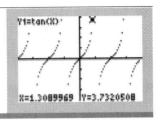

Tip: You probably want to change the **MODE** back from **Dot** to **Connected**.

The Identity $\cos^2 x + \sin^2 x = 1$

[269]   The identity $\cos^2 x + \sin^2 x = 1$ uses a notation that is not even possible to enter on a TI-83. Historically, this is shorthand, meaning $\cos^2 x = (\cos x)^2$. We must use parentheses for $\cos(x)$, and the squaring notation is entered after the closing parenthesis, not before it. See the examples below.

The expression $\cos^2 x$ can be written in any of the following ways: $\cos(X)^2$, $(\cos(X))^2$, or $\cos(X)^2$. The left side of the identity is written as Y₃.

We look at table values starting a 0 and increasing by 0.1. With a good eye for numbers, you can see the pattern that is displayed in Y₃, which is not yet shown.

Arrow over to Y₃. There it is! $\cos^2 x + \sin^2 x = 1$. This important identity will be graphed in Section 7.3.

Reciprocal Functions

[271]   Each trigonometric function introduced so far has a reciprocal function,

$$\frac{1}{\cos(x)} = \text{secant}(x), \quad \frac{1}{\sin(x)} = \text{cosecant}(x), \text{ and } \frac{1}{\tan(x)} = \text{cotangent}(x),$$

but there are no secant, cosecant or cotangent keys on the TI. To use secant in a graph or calculation, you must write it as $1/\cos(x)$.

Enter the two functions. The style has been set to make the basic cosine graph thicker to distinguish the graphs. Use **ZTrig**.

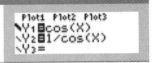

Notice the connecting problem for Y₂. The 1/cos(x) is undefined when the cos(x) = 0. This is precisely where there is a discontinuity.

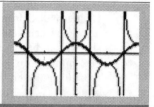

Tip: The COS⁻¹ key on your calculator is not the secant function. Cos⁻¹(x) ≠ sec(x). This key is the inverse cosine and will be introduced in the next section.

6.7 Inverse Trigonometric Functions

In Chapter 4 we encountered the logarithm as the inverse of the exponential function. The logarithm was used to solve exponential equations. In a similar fashion, we now introduce the inverse sine that can be used to solve sine equations. The difference here is that trigonometric functions are periodic, so, for a given output, there will be an infinite number of possible inputs that correspond to a given output value. To be a function, the correspondence must be unique, so we resolve this by restricting the interval of definition.

Solving Using the Inverse Cosine

[274]   Suppose we have a rabbit population modeled by the function $R = -5000\cos(\dfrac{\pi}{6}t) + 10000$. When will the population reach 12,000?

We start with an equation and solve for t

$$12000 = -5000\cos(\dfrac{\pi}{6}t) + 10000$$

$$-0.4 = \cos(\dfrac{\pi}{6}t)$$

$$t = \dfrac{6}{\pi}\cos^{-1}(-0.4).$$

We use the 2ⁿᵈCOS⁻¹ key to enter the expression.

```
(6/π)cos⁻¹(-.4)
         3.785939283
```

We can also graph and use the intersection feature. The Y₂=12000 will give us a horizontal line that is always 12000.

```
Plot1 Plot2 Plot3
\Y1■-5000cos((π/
6)X)+10000
\Y2■12000
\Y3=
\Y4=
\Y5=
\Y6=
```

Use these values as a good viewing window.

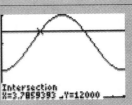

This screen is obtained after the `First Curve`, `Second Curve`, and `Guess` prompting sequence. We used bounds that gave us the intersection to the left. Graphing has an advantage over blindly using `cos`⁻¹ because we can see that there are two values of t where $R=12000$.

The second value of t is found by using symmetry about the vertical line $x = 6$. The second value is to the left of 12, the same distance the first value was to the right of 0. We have $12 - 3.7859393 \approx 8.214$.

> **Tip:**   In mathematical writing the notation arcsin x and sin⁻¹x are interchangeable. The arcsin x notation is not permitted on the TI.

> **Tip:**   Trigonometric functions can be used freely in `solve` and `Solver`.

Graphs of the Inverse Trigonometric Functions

We now graph the three main inverse trigonometric functions. In each case a restriction to the domain of the original function has been made. We start with the $y = \cos^{-1}(x)$; note that this is not the same as the function $y = 1/\cos(x)$, which we call the secant.

Define `Y1=cos`⁻¹`(X)` and graph in a `ZTrig` window to get a first look. We traced to explore the left endpoint on the graph. The graph is quite inadequate because the function is undefined over much of the screen.

The range of the cosine function is the domain of the inverse cosine, so we set $-1 \le x \le 1$. Also, the restricted domain of the cosine function is the range of the inverse function, so we set $0 \le y \le \pi$. The graph exactly fits the screen.

Using the same technique on $y = \sin^{-1}(x)$, we graph with the window set as $-1 \le x \le 1$ and $-\pi/2 \le y \le \pi/2$.

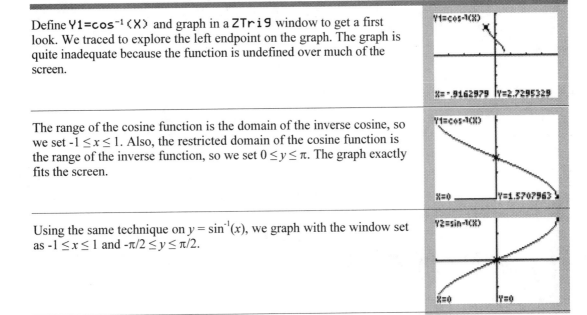

Since the range of the tangent function is all real numbers, the domain of the inverse tangent cannot be fully seen in a window.

We set $-10 \leq x \leq 10$ and use $-\pi/2 \leq y \leq \pi/2$.

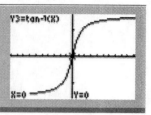

CHAPTER SEVEN

TRIGONOMETRY

7.1 Laws of Sines and Cosines

In many triangle applications a right triangle is involved. In this section we look at solving triangle problems in general.

Law of Cosines with Solver

Solver (or solve) is the handiest way to apply the Law of Cosines. However, if you have since used Solver with another equation, then you must rewrite the equation. We avoid this problem by storing the two laws in the bottom of the Y= menu. Since Y9 and Y0 are rarely used, they are good storage spaces for our formulas. The TI does not have upper and lower case variable names, so we name the sides A, B, and C and the corresponding angles X, Y, and Z. This correspondence must be remembered.

[297] A person leaves her home and walks 5 miles due east and then 3 miles northeast. How far away from home is she?

We enter the formulas in Y9 and Y0 as shown. After entry, the functions have been deselected so that they won't be graphed. (We want to leave them there forever, but we don't want them appearing in any graphs.) Note again that we have used the A-B-C for sides and X-Y-Z for corresponding angles when naming variables.	Plot1 Plot2 Plot3 \Y6= \Y7= \Y8= \Y9=sin(X)/A-sin (Y)/B \Y0=C²-A²-B²+2A* B*cos(Z)
Set the angle mode to Degree. (Not shown.) Use MATH Solver and arrow up to enter Y0, which has the Law of Cosines stored. (Y0 requires a paste from VARS Y-VARS.) Press ENTER.	EQUATION SOLVER eqn:0=Y0∎
Fill in the known variables. Arrow up to C, and make it a guess like 8. Press ALPHA SOLVE to calculate C. The distance from home is about 7.43 miles, but the distance walked is 5+3 = 8 miles.	Y0=0 •C=7.4305587566… A=5 B=3 Z=135 bound={-1ᴇ99,1… •left-rt=0

> *Tip:* In the above example, you could avoid worrying about the angle mode by entering Z=135°. However, if you solve for Z, then the answer is given in the units corresponding to the angle mode setting.

Law of Sines

[299]   Solve the ambiguous case of a triangle with sides $a = 8$, $b = 7$, and $\angle B = 40°$. If you draw a diagram, you see that $\angle A$ can be either acute ($< 90°$) or obtuse ($> 90°$).

Before starting this example, it is noted that we use a different pasting technique than was used in the previous **Solver** example. Choose the way you like best. Start by clearing the **Solver** equation. Press 2nd RCL, and **Rcl** appears at the bottom of the screen as shown. (The RCL is above the **STO▸** key.)	``` EQUATION SOLVER eqn:0= Rcl ```
Now use **VARS Y-VARS** to paste **Y₉** at the bottom of the screen. Press ENTER and the equation itself is pasted into **Solver**. The previous method only showed **eqn:0=Y₉**. The current method shows the actual formula and is thus safer.	``` EQUATION SOLVER eqn:0= Rcl Y₉ ```
Now you explicitly see the Law of Sines, not the **Y₉** variable. Remember our convention that the sides are **A-B-C** and the corresponding angles are **X-Y-Z**. (Sadly, this can't be avoided.)	``` EQUATION SOLVER eqn:0=sin(X)/A-s in(Y)/B ```
We solve for the acute angle first. Input **A=8** for side a, **Y=40** for angle B, and **B=7** for side b. With angles in degrees, we find $\angle A = X \approx 47.3°$.	``` sin(X)/A-sin(…=0 ▪X=47.274644380… A=8 Y=40 B=7 bound={-1ᴇ99,1… ▪left-rt=0 ```
Now reset **X** so that the initial guess is an obtuse angle. Notice how the black squares on the left disappear as an indication that this is no longer a calculated answer. In other words, don't believe the current **X-A-Y-B** values are a solution.	``` sin(X)/A-sin(…=0 X=100▪ A=8 Y=40 B=7 bound={-1ᴇ99,1… left-rt=0 ```
The calculated answer shows $\angle A = X \approx 132.7°$.	``` sin(X)/A-sin(…=0 ▪X=132.72535561… A=8 Y=40 B=7 bound={-1ᴇ99,1… ▪left-rt=0 ```

> *Tip:*   The two laws can be left in **Y₉** and **Y₀** then brought back whenever they are needed. Just be sure that they are deselected (turned off) so that they do not affect graphing.

7.2 Trigonometric Identities

Identities are often used to simplify or to factor equations that are being solved. The TI is not of much use in this regard, but it does offer graphical techniques that can replace the need to solve algebraically. If answers need only be approximations, a graphical method is acceptable. In cases where the answer must be exact, the graphical approach is an excellent means of testing algebraically derived results.

[302]   Show the identity $\sin^2 t + \cos^2 t = 1$ graphically.

This is a fun surprise. Enter the left-hand side of the identity and graph in a **ZDecimal** window. The graph is clearly the constant function $y = 1$, which graphically verifies the identity. In Section 6.6 we also looked at this identity as a table.

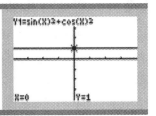

[304]   Find all solutions to $\sin 2t = \sin t$, for $0 \leq t \leq 2\pi$.

Change to **Radian** angle mode. Enter $Y_1 = \sin(2X)$ and $Y_2 = \sin(X)$. Make the following **WINDOW** settings.

Press **TRACE** and you see the cursor in the middle of the screen. Luckily it is at one of the intersections. Applying common sense to the bottom line information ($x \approx \pi$ and $y \approx 0$) tells us that one of the solutions is $t = \pi$.

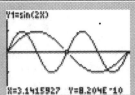

We can use the **2nd CALC intersect** to find the next intersection to the right. These values can be converted to exact values by noting that $y = -\sqrt{3}/2$, which means $x = 5\pi/3$. Knowing symmetry we can list all the solutions:

$$t = 0, \ \pi/3, \ \pi, \ 5\pi/3, \ 2\pi.$$

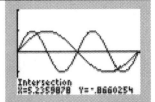

[305]   Show the identity $\cos 2t = 1 - 2\sin^2 t$ graphically.

Enter the two functions as shown. The thick style is used for Y_2, so that we see visually that the second function is the same graph as the first.

We see a graphical verification that the equation is an identity. In this screen, the thick graph is seen as it is writing over the first graph.

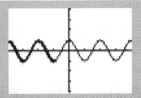

Tip: Graphic verification is only a common-sense indicator. Formal proof requires algebra.

7.3 Sum and Difference Formulas for Sine and Cosine

Two of the most important trigonometric identities are

$$\sin(\theta + \phi) = \sin\theta\cos\phi + \sin\phi\cos\theta$$

$$\cos(\theta + \phi) = \cos\theta\cos\phi - \sin\phi\sin\theta.$$

By using these and the Pythagorean identity, numerous other identities can easily be derived. For example, let ϕ be negative and you have the difference formulas. Or let $\theta = \phi$ and you have the double-angle formulas. There are many more examples.

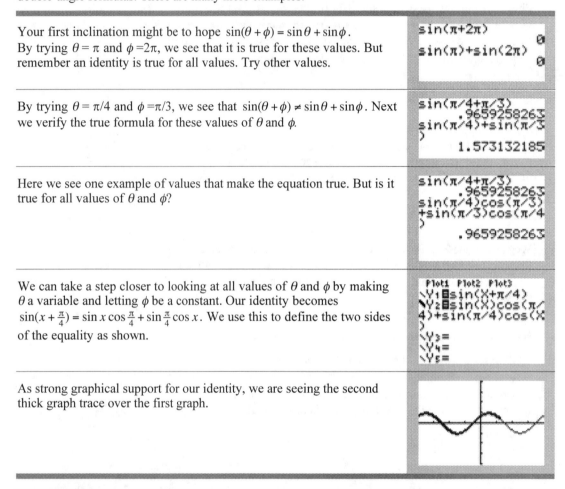

Your first inclination might be to hope $\sin(\theta + \phi) = \sin\theta + \sin\phi$. By trying $\theta = \pi$ and $\phi = 2\pi$, we see that it is true for these values. But remember an identity is true for all values. Try other values.

By trying $\theta = \pi/4$ and $\phi = \pi/3$, we see that $\sin(\theta + \phi) \neq \sin\theta + \sin\phi$. Next we verify the true formula for these values of θ and ϕ.

Here we see one example of values that make the equation true. But is it true for all values of θ and ϕ?

We can take a step closer to looking at all values of θ and ϕ by making θ a variable and letting ϕ be a constant. Our identity becomes $\sin(x + \frac{\pi}{4}) = \sin x \cos\frac{\pi}{4} + \sin\frac{\pi}{4}\cos x$. We use this to define the two sides of the equality as shown.

As strong graphical support for our identity, we are seeing the second thick graph trace over the first graph.

Tip: For identity checking, the graphing style ◆ is also a good choice.

Rewriting sin(t) + cos(t)

[309]   The previous graph showed that the Y2 function, which is essentially a sum of a sine and a cosine function, is equal to Y1, a single sine function. Using the sum identity and special angle values,

$$\sin(x)\cos(\pi/4) + \sin(\pi/4)\cos(x) = (\sqrt{2}/2)\sin(x) + (\sqrt{2}/2)\cos(x) = \sin(x + \pi/4).$$

By multiplying both sides by $\sqrt{2}$ we can write

$$\sqrt{2}\sin(x + \pi/4) = \sin(x) + \cos(x).$$

Enter the function $Y_1 = \sin(X) + \cos(X)$ and graph in a ZTrig window. TRACE shows that the Y_1 graph is not at a maximum when $x = 0$ and $y = 1$. The amplitude of Y_1 must be greater than 1, and from $\sqrt{2}\sin(x + \pi/4)$, we know it is actually $\sqrt{2}$. The Y_1 graph looks like a sine curve shifted left by some amount and from $\sqrt{2}\sin(x + \pi/4)$ we know the shift is $\pi/4$.

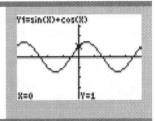

[310] The general textbook result used to write the sum of a sine and a cosine as a single sine is:

$$A\sin(Bx + \phi) = a_1\sin(Bx) + a_2\cos(Bx)$$

where $A = \sqrt{a_1^2 + a_2^2}$ and $\tan\phi = \dfrac{a_2}{a_1}$.

7.4 Trigonometric Models

Dampened Oscillations

[320] A classic physics experiment is to track data on the oscillation of a spring. We can use a trigonometric function to model motion. However, this model may only fit in the short run since we know that the oscillations die out over time. We graph a model that both oscillates and dies out.

Our basic oscillation model is $y = \cos(2\pi x)$, which we dampens by using the exponential function $y = 5(1/2)^x$. The definitions of Y_1 and Y_2 form an upper and lower bound for the function model that is written as Y_3.

```
Plot1 Plot2 Plot3
\Y1■5(1/2)^X
\Y2■-5(1/2)^X
\Y3■5(1/2)^X*cos
(2πX)
\Y4=
\Y5=
\Y6=
```

This window fits our graph.

```
WINDOW
 Xmin=0
 Xmax=5
 Xscl=1
 Ymin=-5
 Ymax=5
 Yscl=1
 Xres=1
```

Here we see the oscillating graph Y_3 with the two exponential functions Y_1 and Y_2 serving as an envelope.

Near $x = 5$, in the above graph it looks as though the wiggling has stopped. We now reset the window to see the next interval of length 5, the interval $5 \leq x \leq 10$. The y-values are much smaller; they have been specially chosen to make this graph look like the one before.

```
WINDOW
 Xmin=5
 Xmax=10
 Xscl=1
 Ymin=-.15
 Ymax=.15
 Yscl=1
 Xres=1
```

The graph looks like the one on the interval $0 \le x \le 5$. We only know we are seeing a new graph because of the trace values shown at the bottom of the screen. The oscillation continues forever in this model, but y-values become very small as x gets large.

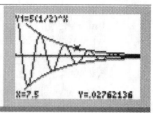

7.5 Polar Coordinates

This section first introduces the procedure needed to convert between our standard Cartesian coordinate system and a polar coordinate system. (The text uses the term Cartesian, while the TI calls them rectangular coordinates.) Second, we look at the techniques for graphing a function that is defined in polar variables. We will use only `Radian` angle mode in this section.

Converting Coordinates

The polar coordinate of a point P in the xy-plane is a pair r and θ, where r is the direct distance from P to the origin and θ is the angle measured counterclockwise from the positive x-axis to the line joining P and the origin.

[326]   Give Cartesian coordinates for the polar coordinate point $(7, \pi/3)$.
Give polar coordinates for the Cartesian coordinate point $(3, 4)$.

Conversion takes place on the home screen. We first press 2nd`ANGLE` to see the conversion items; they are listed as `5:R▸Pr(` through `8:P▸Ry(`. They are also found in the `CATALOG`.	`ANGLE`
	`1:°`
	`2:'`
	`3:r`
	`4:▸DMS`
	`5:R▸Pr(`
	`6:R▸Pθ(`
	`7:P▸Rx(`

To get the x-coordinate, we paste the `P▸Rx(` on the home screen and complete the command with the polar coordinates as shown. To save some redundant entry for the y-coordinate, you can use 2nd`ENTRY` and paste `P▸Ry(` over `P▸Rx(`.	`P▸Rx(7,π/3)`
	`                3.5`
	`P▸Ry(7,π/3)`
	`       6.062177826`

We continue and convert $(3, 4)$ from rectangular coordinates to polar coordinates, obtaining the r-value and the θ-value, as shown.	`                3.5`
	`P▸Ry(7,π/3)`
	`       6.062177826`
	`R▸Pr(3,4)`
	`                  5`
	`R▸Pθ(3,4)`
	`          .927295218`

Tip: In general, the TI uses the symbol ▸ to indicate translation/conversion expressions. We have seen ▸`Frac`, which converts a decimal to a fraction, and ▸`Eff`, which converts a nominal interest rate to an effective interest rate. Now we use `P▸Rx`, which converts a polar coordinate pair to the rectangular x-coordinate.

Tip: If you use the `CATALOG`, press `Q` and the up-arrow key to get to the `P▸R` listings more quickly than pressing `P` and using down arrow. Use `S` and the up-arrow key for the `R▸P` listings. But overall, the `ANGLE` menu seems the best choice.

Graphing Equations in Polar Coordinates

[327]   Next we see how to graph a spiral as a polar equation. The graphing modes (Func Par Pol Seq) have many similarities: The Y= menu keeps the function definitions, the viewing windows are set in WINDOW, TRACE is used to see function values, and TABLE shows lists of numerical values.

We use MODE to change the function mode from Func to Pol.	
Press Y= and see that the function names start with r1=. Polar equations are written as r in terms of θ (instead of y in terms of x). If the function mode is Pol, the variable key X,T,θ,n will enter θ, not X. (You can also use ALPHA 3 to enter θ.)	
When in Pol mode, the window starts with three new θ settings. These control the values plotted. Shown is the ZStandard window, and it says start at θ = 0 and graph points until θ = 2π. Each new point uses the previous θ plus an increment of θstep. (The ZStandard value of θstep is π/24.) The actual screen size is controlled, as before, with the X,Y settings.	
Pressing ZOOM 5:ZStandard and TRACE shows the spiral graph. Press the right-arrow key repeatedly to have the cursor trace the graph. It may seem strange, but from the point shown here, pressing the right-arrow (meaning increase θ) will make you move left on the graph.	
In the previous screen, the bottom-line coordinates given were X, Y, and θ. To see the polar graph coordinates (PolarGC), we use the FORMAT menu. Press 2nd FORMAT (above ZOOM).	
After changing the PolarGC setting, we see R and θ values as we trace.	
Often the θ setting, 0 ≤ θ ≤ 2π, needs to be longer to show more of the graph. To create the graph shown here, we used 0 ≤ θ ≤ 3π. You can see that the spiral continued an extra π or 180° from the previous graph.	

Tip: Many beautiful graphs can be created using polar coordinates. See Chapter 5 in the *TI–Guidebook* for an example.

> *Tip:* When you change from `Func` to `Pol`, the `Y` defined functions are not erased, and the `r=` definitions are shown in their place. Likewise, by changing from `Pol` to `Func`, the `r=` definitions are not erased but will not be seen on the `Y=` menu screen.

> *Tip:* Even in `Func`, you can change to `PolarGC`, so that `TRACE` reports points in polar coordinates.

7.6 Complex Numbers and Polar Coordinates

Definition: A *complex* number is defined as any number that can be written in the form

$$z = a + bi$$

where *a* and *b* are real numbers and $i = \sqrt{-1}$.

> *Tip:* Do not confuse the *i*-vector from Chapter 10, with the number *i*.

Complex Numbers as Solutions to Equations

[331]   Solve $x^2 - 2x + 2 = 0$.

Recall in using the `QF` program, we got an error message in one case because the two calculations were not real numbers.
 	
The program code is shown here as a reminder.
	`PROGRAM:QF`
	`:Prompt A,B,C`
	`:Disp (-B+√(B²-4`
	`A*C))/(2A)`
	`:Disp (-B-√(B²-4`
	`A*C))/(2A)`
	`:`

Set the next to last line in the `MODE` menu from `Real` to `a+b`*i*.	`Normal Sci Eng`
	`Float 0123456789`
	`Radian Degree`
	`Func Par Pol Seq`
	`Connected Dot`
	`Sequential Simul`
	`Real a+bi re^θi`
	`Full Horiz G-T`

When the program is run and the values are input, the output is now two complex numbers instead of an error message.	`prgmQF`
	`A=?1`
	`B=?-2`
	`C=?2`
	`            1+i`
	`            1-i`
	`          Done`

> *Tip:* If you use `solve` or `Solver`, they will not return complex numbers as answers.

Algebra of Complex Numbers

[332]   Write $(2+7i)(4-6i) - i$ as a single complex number. Compute $(2+7i)/(4-6i)$.

We enter the complex numbers as given. (The *i* is a **2nd** key above the period in the bottom row of the keypad.) The first calculation is clear. However, when we do the division, the answer is not exact and not completely visible on the screen. To correct this we press **MATH 1**, and the answer is translated to exact fractional form. (The **Ans** is automatically pasted when you press **MATH 1** and does not need to be entered.)

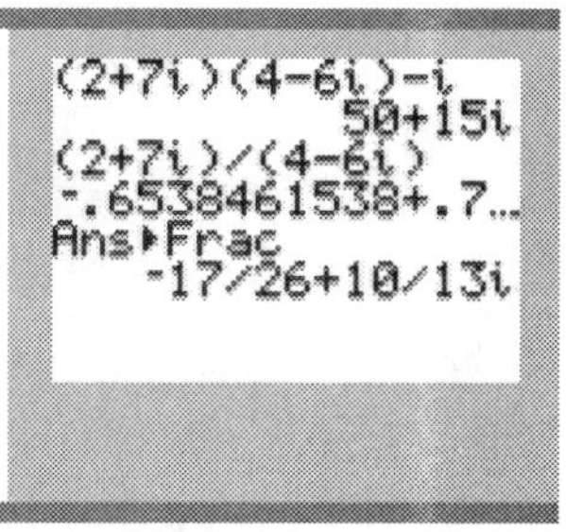

```
(2+7i)(4-6i)-i
            50+15i
(2+7i)/(4-6i)
   -.6538461538+.7…
Ans▶Frac
      -17/26+10/13i
```

[334]   Convert 4^i to the form $a + bi$.

In this example, we see that complex numbers are even permitted as exponents. Because we expect decimal components that run off the screen we change **MODE** to show 3 rounded decimals.

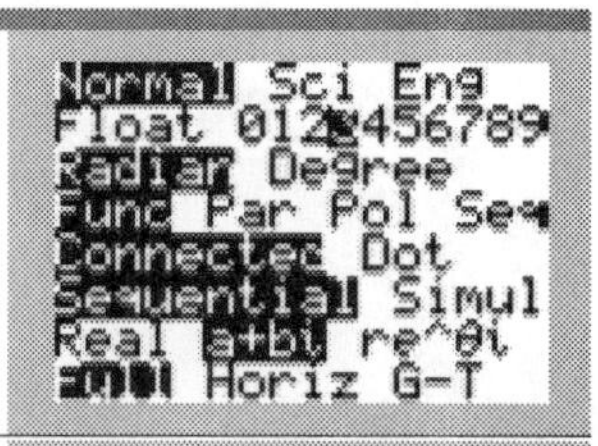

```
Normal Sci Eng
Float 012▊456789
Radian Degree
Func Par Pol Se◄
Connected Dot
Sequential Simul
Real a+bi re^θi
▊▊▊ Horiz G-T
```

Now the calculations stay on the screen.

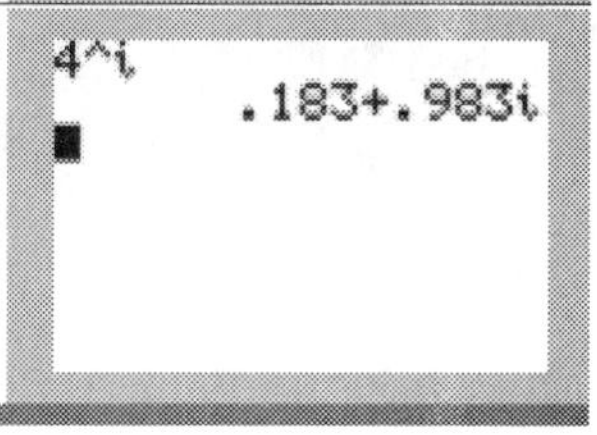

```
4^i
            .183+.983i
■
```

Polar Form of a Complex Number

Polar form is a compact way of expressing a complex number by its polar coordinates. In this subsection, for better screen output, we continue with the **MODE** set to **3**-digit rounding. The angle mode is **Radian**.

[333]   Convert $z = -2i$ and $z = -2 + 3i$ to polar form.
Change the polar coordinates $r = 5$ and $\theta = 3\pi/4$ to complex form $a + bi$.

The **MATH CPX** menu has the two translation options **6: ▶Rect** and **7: ▶Polar**. These are used in the next screen to make the desired conversions.

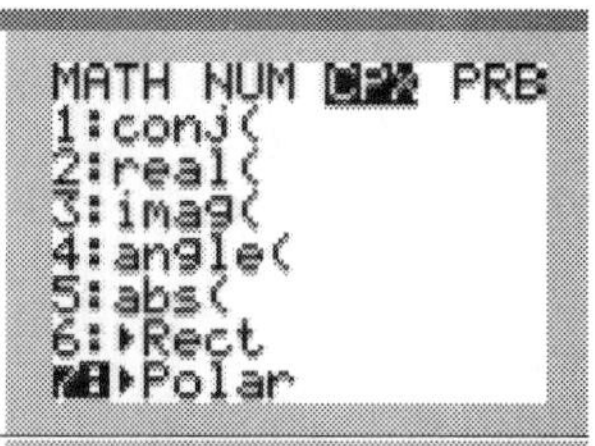

```
MATH NUM ▊▊▊ PRB
1:conj(
2:real(
3:imag(
4:angle(
5:abs(
6:▶Rect
7▊▶Polar
```

Enter complex numbers by using *i*. If you forget the *i* in either form, you will get a **DATA TYPE** error. It is easy to forget in the polar form, so practice using **e^(** *i* together, as shown here.

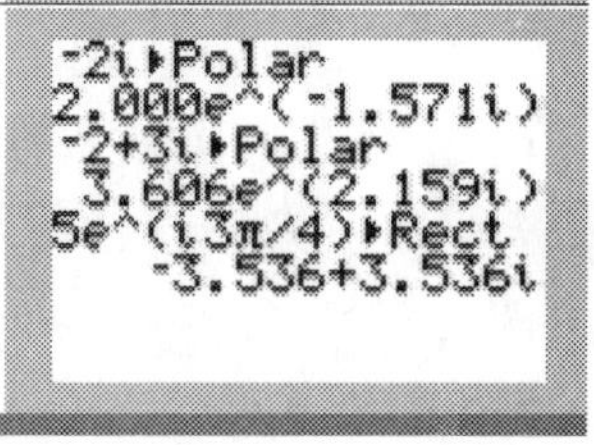

```
-2i▶Polar
2.000e^(-1.571i)
-2+3i▶Polar
3.606e^(2.159i)
5e^(i3π/4)▶Rect
     -3.536+3.536i
```

> **Tip:**   When making frequent changes to **MODE** options, it is a good idea to check the settings before starting on any new problem, since the previous settings might no longer be desirable.

Notes:

CHAPTER EIGHT

COMPOSITION, INVERSES, AND COMBINATIONS

8.1 Composition of Functions

[344]   Let $p(x) = 2x + 1$ and $q(x) = x^2 - 3$. Suppose $u(x) = p(q(x))$ and $v(x) = q(p(x))$.

(a) Calculate $u(3)$ and $v(3)$.

(b) Find formulas for $u(x)$ and $v(x)$.

We enter the functions p and q as Y₁ and Y₂, then use these to enter u and v as Y₃ and Y₄. Since we only want values of u and v, we deselect Y₁ and Y₂.

Use 2ⁿᵈ TBLSET to start at 0 with ▵Tbl=1. The table will then include the values $u(3) = 13$ and $v(3) = 46$.

For part (b), the TI-83 cannot give an algebraic formula, but you can check the graph of your algebraic derivation against the graph of Y₃ and Y₄.

Algebraically, we derive: $u(x) = 2x^2 - 5$ and $v(x) = 4x^2 + 4x - 2$. The Y₃ graph is seen as $y = 2x^2$ shifted down by 5, which makes sense from the formula. The Y₄ graph could be more easily checked if v was in vertex form, but we easily see that the y-intercept is –2.

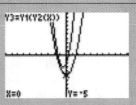

Tip: You can shorten a composition definition by not showing the variable X. For example, Y₃ = Y₁(Y₂), but showing X is closer to function composition symbolism.

8.2 Inverse Functions

In Section 6.7, we introduced the inverse sine function $y = \sin^{-1}(x)$. The equation $\sin(x) = 0.8$ can be solved by applying the inverse to both sides and using the TI to find $x = \sin^{-1}(0.8) \approx 0.93$. In this section, we look at inverse functions in general. If $y = f(x)$, then $x = f^{-1}(y)$. Inverse functions are functions in their own right and need not use the $^{-1}$ notation. For example, if $y = f(x) = 10^x$, then the inverse function is $y = \log(x)$. Not all functions have an inverse; those that do are called *invertible*.

An Invertible Function

[354]   Let $P(x) = 2^x$.
   (a) Show that P is invertible.
   (b) Graph P and P^{-1}.

We graph the function and view it in a ZDecimal window. We see that the function is always increasing and that any horizontal line would pass through the graph in at most one place. The function is invertible.	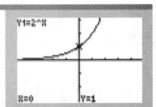
The graphs of a function and its inverse are symmetrical across the line $y = x$. The TI will draw an inverse graph without knowing its formula. Press 2ⁿᵈDRAW, and select 8:DrawInv. The expression DrawInv will be pasted to the home screen.	
Finish this command by pasting the function name Y₁.	
The graphical inverse is drawn. There is a drawback in that you cannot trace a "drawn" inverse. It is just a graphical object drawn on the existing screen. But, as shown here, you can press the arrow keys and move the cross-hair cursor to the graph. This shows approximate x and y values.	

Tip: You can algebraically find the inverse of $P(x) = 2^x$, as $P^{-1}(x) = \log(x) / \log(2)$ then graph $P^{-1}(x)$ to see that it matches the inverse function drawn above.

Tip: When pasting commands such as DrawInv from a graph window, be sure to check that it is pasted on a new line when you return to the home screen. If not, the pasted command continues on the command line and almost certainly causes a syntax error.

Checking Inverse Formulas

[356]   Check that $f(x) = \dfrac{x}{2x+1}$ and $f^{-1}(x) = \dfrac{x}{1-2x}$ are inverse functions of one another.

We start by entering the functions as Y₁ and Y₂. Set Y₂ to thick style. Use ZDecimal. Shown here is Y₁ being traced at the point $x = -0.5$. This point is not in the domain of Y₁ since it has no y-value as shown. We strongly suspect that the two functions are inverses because they appear symmetric about the line $y = x$. (The symmetry line is not shown.)	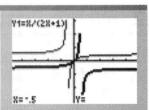

A clever way to further test inverse formulas is to compose them. We know for inverse functions: $f(f^{-1}(x)) = x$ and $f^{-1}(f(x)) = x$. We write these as Y₃ and Y₄.

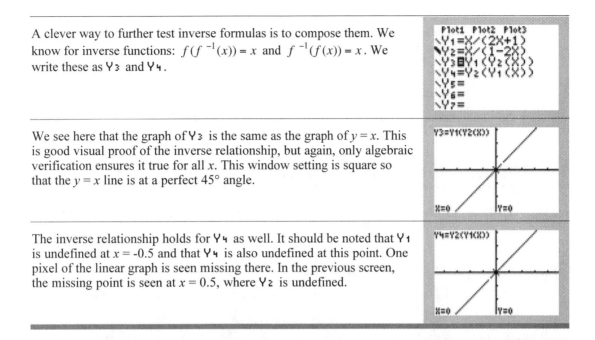

We see here that the graph of Y₃ is the same as the graph of $y = x$. This is good visual proof of the inverse relationship, but again, only algebraic verification ensures it true for all x. This window setting is square so that the $y = x$ line is at a perfect 45° angle.

The inverse relationship holds for Y₄ as well. It should be noted that Y₁ is undefined at $x = -0.5$ and that Y₄ is also undefined at this point. One pixel of the linear graph is seen missing there. In the previous screen, the missing point is seen at $x = 0.5$, where Y₂ is undefined.

Tip: The missing points on the two linear graphs showed because the window setting `ZDecimal` caused x to be evaluated there. Other settings may not evaluate x and not show the undefined point.

Restricting Domains for Inverses

[358]   We have seen that the sine function must be restricted to the interval $-\pi/2 \leq x \leq \pi/2$ for it to have an inverse. We graph the full sine function and use `DrawInv` to see why the restriction is necessary.

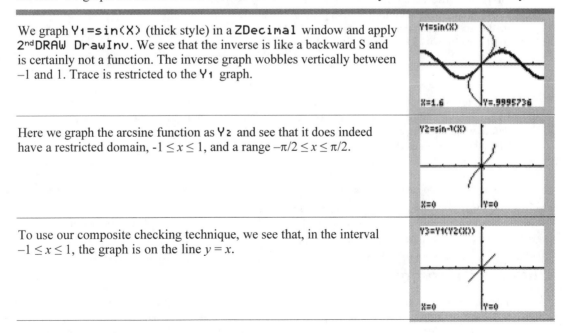

We graph Y₁=sin(X) (thick style) in a `ZDecimal` window and apply 2ⁿᵈDRAW DrawInv. We see that the inverse is like a backward S and is certainly not a function. The inverse graph wobbles vertically between -1 and 1. Trace is restricted to the Y₁ graph.

Here we graph the arcsine function as Y₂ and see that it does indeed have a restricted domain, $-1 \leq x \leq 1$, and a range $-\pi/2 \leq x \leq \pi/2$.

To use our composite checking technique, we see that, in the interval $-1 \leq x \leq 1$, the graph is on the line $y = x$.

Just for fun, instead of Y₃=Y₁(Y₂(X)) you might check the other inverse relationship Y₄=Y₂(Y₁(X)). What is happening here? This is not the line $y = x$! However, remember that the function Y₁ is restricted to the interval $-\pi/2 \leq x \leq \pi/2$. On that interval, the graph coincides with the graph of $y = x$.

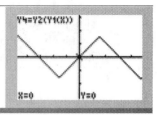

8.3 Combinations of Functions

We used combinations of functions, when we defined a sum and a quotient for trigonometric functions in Section 6.6. Now we look in general.

Graphs of Combinations

[361]   In Section 4.2 we introduced the population versus food model proposed by Malthus.

$$P = 2(1.04)^t \text{ and } N = 4 + 0.5t.$$

We now look at two ways to combine these functions to measure prosperity.

In Y₃ the difference of the two functions is formed. When Y₃ is zero, the food supply is the same as the population. After that point in time, a shortage occurs.

Graph the difference function with a window $0 \leq x \leq 100$ and $-5 \leq y \leq 20$. We traced to get an approximation of the maximum difference, which occurs at about year 47. At this point there is enough food for almost 15 million additional people.

Now graph the quotient function Y₄ with a window $0 \leq x \leq 100$ and $-1 \leq y \leq 5$. We traced to get an approximation of the maximum quotient value, which occurs at about year 17. At this point there is enough food to feed over 3 times that population.

Thus, the two methods of measure, difference and quotient, give different answers for the time of maximum prosperity. However, to find when the food shortage begins, either graph yields the same answer, approximately 78 years.

Advanced Uses of Lists

The TI-83 has the ability to name lists in addition to the standard lists, L1 to L6. We limit ourselves to looking at how to store a list by naming it and how to get it back as one of the L1 to L6 lists. Interested readers can find a fuller presentation of list features in Chapters 11 and 12 of the *TI-Guidebook*.

Naming a list

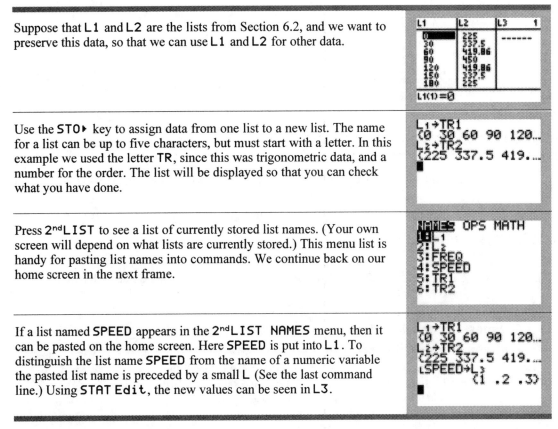

Suppose that **L1** and **L2** are the lists from Section 6.2, and we want to preserve this data, so that we can use **L1** and **L2** for other data.

Use the **STO▸** key to assign data from one list to a new list. The name for a list can be up to five characters, but must start with a letter. In this example we used the letter **TR**, since this was trigonometric data, and a number for the order. The list will be displayed so that you can check what you have done.

Press **2ⁿᵈLIST** to see a list of currently stored list names. (Your own screen will depend on what lists are currently stored.) This menu list is handy for pasting list names into commands. We continue back on our home screen in the next frame.

If a list named **SPEED** appears in the **2ⁿᵈLIST NAMES** menu, then it can be pasted on the home screen. Here **SPEED** is put into **L1**. To distinguish the list name **SPEED** from the name of a numeric variable the pasted list name is preceded by a small **L** (See the last command line.) Using **STAT Edit**, the new values can be seen in **L3**.

In summary, we have swapped list of data in and out of a standard list name. You are allowed up to 20 list names to show in **STAT Edit** and they can be arranged in any order. But we will use a simplified approach and view only the conventional six lists, **L1** to **L6**. See Chapter 11 and 12 of the *TI–Guidebook* for details on list arrangements.

Notes on deleting lists and other stored items with a TI-83 Plus

Press **2ⁿᵈMEM** to see a **MEMORY** menu of options. To clear all the lists at once, you can select **4:ClrAllLists**. For this example we will delete individual lists by using the option **2:Delete**.

Next, we choose **List** as the kind of object to delete.

When the arrow (▸) is to the left of an item, it is ready to be deleted. Press **DEL** to delete. Be careful about your choice because once deleted the list is gone forever. If there is any doubt about what to do, use **2ⁿᵈQUIT** to get out.

A safer way to delete is by selecting one or more lists to be deleted. Select a list name by pressing ENTER. An asterisk appears beside the name. The asterisk is turned both on and off by ENTER.

With the appropriate names selected press DEL. You will be asked before the delete.

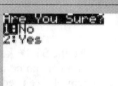

Tip: If you delete an L-list, it is also be deleted from the STAT Edit menu if it was showing there.

Tip To restore the standard list order, press: STAT 5:SetUpEditor ENTER. This resets the six standard lists.

Notes on deleting lists and other stored items with a non-Plus TI-83

Press 2ndMEM to see a MEMORY menu of options. To clear all the lists at once, you can select 4:ClrAllLists. For this example we will delete individual lists by using the option 2:Delete.

Next, we choose List as the kind of object to delete.

When the arrow (▸) is to the left of an item, it is ready to be deleted. Press ENTER to delete. Be careful about your choice because once deleted the list is gone forever. If there is any doubt about what to do, use 2ndQUIT to get out.

Combinations of Functions Defined by Tables

[366] In this section, we look at two functions defined by data to form a quotient that represents a per capita crime rate.

Crime table

Year since 1997	0	1	2	3	4	5
Crime, City A	793	795	807	818	825	831
Crime, City B	448	500	525	566	593	652

Population table

Years since 1997	0	1	2	3	4	5
Population, City A	61000	62100	63220	64350	65510	66690
Population, City B	28000	28588	29188	29801	30427	31066

We enter the crime data in $L1$ and $L2$ and the population data in $L3$ and $L4$.

We will use $L5$ and $L6$ to calculate the ratios that represent the per capita crime rate. Move the cursor to highlight $L5$ and press **ENTER**. Enter the quotient formula as shown.

The previous technique could be repeated, but we show another approach and store a formula in the header. In this screen you see that by enclosing the formula in quotes (**"**) the formula itself is both applied and stored. A special symbol (◈) appears in the header to alert you that it contains a formula. Move the cursor up to $L6$ to see the formula.

We see that the per capita crime rate for City A starts at 1.3% and falls over the next five years, whereas the per capita crime rate for City B starts at 1.6% and rises over the same period.

Tip: To delete a list formula, enter a blank formula (for example, $L6=$ **" "**).

Notes:

CHAPTER NINE

POLYNOMIAL AND RATIONAL FUNCTIONS

9.1 Power Functions

Definition: A quantity y is proportional to a power of x if

$$y = kx^n,$$

where k and n are constant.

Directly Proportional

[376]   We compare three functions — the circumference of a circle, the area of a circle, and the volume of a sphere — each written in terms of its radius. These are all examples of a direct proportion.

$$C = 2\pi r \qquad A = \pi r^2 \qquad V = \frac{4}{3}\pi r^3$$

Graphs

Use Y= to show the formula screen. If you need to delete a previous formula, press the CLEAR key, with the cursor on the formula line. Here we show the three proportional functions that have been entered. Note that the = signs are highlighted to show these formulas are active.

```
Plot1 Plot2 Plot3
\Y1◼2πX
\Y2◼πX²
\Y3◼(4/3)πX^3
\Y4=
\Y5=
\Y6=
\Y7=
```

We set graph style to distinguish between the three graphs. At the far left of each function definition is a graph style indicator. Arrow over and repeatedly press ENTER to change the style. We use regular, thick, and dot as shown.

```
Plot1 Plot2 Plot3
\Y1◼2πX
\Y2◼πX²
∙Y3◼(4/3)πX^3
\Y4=
\Y5=
\Y6=
\Y7=
```

Use WINDOW to set the viewing window values as shown.

```
WINDOW
 Xmin=0
 Xmax=6
 Xscl=1
 Ymin=0
 Ymax=100
 Yscl=10
 Xres=1
```

Press **TRACE** to see the graph and be ready to trace.

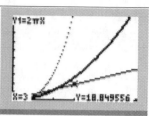

Inversely Proportional

[376] The weight, w, of an object is inversely proportional to the square of the objects distance, d, from the earth's center:

$$w = \frac{k}{d^2} = kd^{-2}$$

Make a graph and table for the weight an object in terms of its distance from earth's center. On the surface of the earth, which is 3959 miles from the earth's center, it weighs 44 pounds.

| Use the formula to find k from the values we know. | `44*3959²→K` `         689641964` |

To make it more interesting we define the function with a fraction in **Y₄** and with a negative exponent in **Y₅**

```
Plot1 Plot2 Plot3
\Y1=2πX
\Y2=πX^2
\Y3=(4/3)X^3
\Y4◼K/X^2
\Y5◼K*X^(-2)◼
\Y6=
\Y7=
```

Before pressing **TRACE**, set the viewing window to $0 \le x \le 10000$ (Xscl=4000) and $0 \le y \le 110$ (Yscl=50). Notice that there appears to be only one graph; use the up arrow to toggle between graphs and confirm that the graphs are indeed the same.

Check the table to further confirm that the definitions are the same. Press **2ⁿᵈTBLSET** and set **TblStart=0** and **△Tbl**=1000. Press **2ⁿᵈTABLE** to see the table shown. Notice that the two functions have the same output values. In addition you should note **ERROR** on the top line. In this case, it tells you that division by 0 is an error.

X	Y₄	Y₅
0	ERROR	ERROR
1000	689.64	689.64
2000	172.41	172.41
3000	76.627	76.627
4000	43.103	43.103
5000	27.586	27.586
6000	19.157	19.157

X=0

> **Tip:** Use common sense when interpreting a table with identical values or a graph that is the same in the viewing window. It may be a local phenomenon.

Definition: A *power* function is a function of the form

$$f(x) = kx^p, \text{ where } k \text{ and } p \text{ are constant.}$$

[378] You can explore various effects that k and p have on the family of power functions. For example, graph the four functions $y = x^{-1}, x^{-2}, x^{-3}, x^{-4}$, which you enter as **Y₁** to **Y₄**, or as one function using

set notation, Y₁=X^{-1, -2, -3, -4}. Graph using a ZDecimal window to get a first view. The graph (not shown) shows a similarity between the even negative power functions and another similarity between the odd power functions.

To show the difference of behavior, we investigate the graph of one odd representative, Y1=X^-1, and one even representative, Y2=X^-2. For negative *x*, the odd functions will be negative, and the even functions will be positive.

If you are investigating the behavior of function values as they get close to a point (often zero), then first set Indpnt: Ask in the TBLSET menu. For example, after Ask is set, the table shown can be constructed. The *x*-values approach zero from the positive side. No setting of ∆Tbl could generate this table.

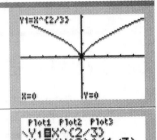

Note on fractional power graphs

Fractional powers, such as $x^{2/3}$ are acceptable, but they should be entered with parentheses to ensure proper meaning.

We enter the function Y1=X^(2/3) and graph with a ZDecimal window.

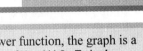

These three definitions are the same.

Tip: If you forget to use parentheses when entering a fractional power function, the graph is a power function of the numerator only. For example, the graph of Y1=X^2/3, is the parabola $y = (1/3)x^2$.

9.2 Polynomials

This section looks at polynomials. A polynomial is a sum of power functions whose exponents are nonnegative integers. The two issues addressed here are how to get a good graphing window and how to find the zeros of a polynomial.

Graphs of Polynomials

[387]   Show that the graph of $f(x) = x^3 + x^2$ resembles the power function $y = x^3$ on a large scale.

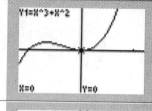

Enter Y₁=X^3+X^2 and use ZDecimal to get a first impression. The behavior of the function near the origin is not shown very well. To get a better view we zoom in. Press ZOOM ZoomIn ENTER, then TRACE, to see the screen shown. Now we see the characteristic curve of a polynomial.

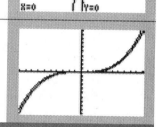

In the long-run, a polynomial looks like the graph of $y = x^2$ or $y = x^3$, depending on its leading term. To see this behavior, we zoom out twice, once to get back to ZDecimal and once more to get the long-run view shown by this screen.

Press ZOOM ZoomOut ENTER ENTER, then TRACE.

The previous window showed us that this function looks like $y = x^3$ in the long-run. To get a better feel for this, you would need to show more of the y-axis. If you enter Y₂=X^3 and reset the window to $-10 \le x \le 10$ and $-1000 \le y \le 1000$, then the graph of Y₁ is close to the graph of Y₂.

Zeros of Polynomials

[387]   Given the polynomial $q(x) = 3x^6 - 2x^5 + 4x^2 - 1$, find the intercepts. That is, where is $q(0)$ and for what x is $q(x) = 0$?

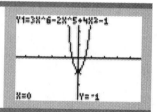

Enter the function and use a ZDecimal window. We can see immediately that the y-intercept is given by $q(0) = -1$. There are two x-intercepts at $x \approx \pm 0.5$. The x-intercepts are called zeros and will be discussed in the next section.

9.3 Short-Run Behavior of Polynomials

We zoomed in and zoomed out on the graph of a function to see the short-run and long-run behavior of the polynomial. In many applications where polynomial functions are used as models, we need to find the zeros of the function. In the previous section, we have seen that the zeros of a function are where the graph crosses the x-axis.

Graphs Showing Polynomial Zeros

[390]   Investigate the short-run behavior of $u(x) = x^3 - x^2 - 6x$. Where does the graph cross the x-axis?

Enter Y1=X^3-X^2-6X and use ZDecimal to get a first impression. This is a poor view because we see three pieces of the graph. You may want to zoom out for a better view. However, we do see all three zeros in this window, so we estimate them from this window as $x = -2, 0, 3$. You can evaluate Y1 to check that Y1(-2)=0, Y1(0)=0, and Y1(3)=0.

These zeros are quite obvious in this case. If the values are not obvious, you can find them by using 2nd CALC 2:zero.

The same procedure used for minimum, or maximum, is used at this point. You are prompted for

Left Bound?, Right Bound?, Guess?.

With these suitably chosen, you can find the largest zero is at $x = 3$.

Multiple Zeros

[392]   Describe the graphs of $y = (x+1)^2$, $y = (x+1)^3$, $y = (x+1)^4$, $y = (x+1)^5$, and so on.

This graph can be seen as having a zero at $x = -1$. The point of interest is that this differs from the zeros in the previous example because the graph does not continue through the x-axis. In nonmathematical terms, it just stops by for a kiss.

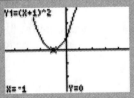

In the next case, you can see that the graph *does* cross the x-axis. However, by close inspection, you can also see that the graph is quite flat near the zero.

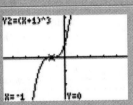

Next we are back to a graph similar to the first graph. We can see a pattern that even powers will look like a $\cup$ that touches the x-axis.

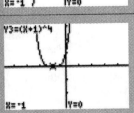

Finally, we conclude that odd powers pass through the x-axis but are quite flat at the zero. Since the function can be factored as

$$y = (x+1)^5 = (x+1)(x+1)(x+1)(x+1)(x+1),$$

we see that $x = -1$ is a zero five times. It is a multiple zero.

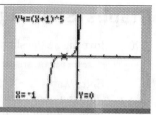

9.4 Rational Functions

Definition: A *rational* function is one that can be written as a ratio of two polynomials. (We assume that the denominator is not the constant 0.) This definition means that even if a function is not in the form of a numerator over a denominator, it could still be considered rational if it could be written that way. For example,

$$f(x) = x + 3 - \frac{2}{x-2} \quad \text{can also be written as} \quad f(x) = \frac{x^2 + x - 8}{x-2}.$$

The Long-Run Behavior

The long-run behavior of a rational function in fraction form is determined by the quotient of the highest power terms in the numerator and denominator.

[397] Show that the long-run behavior of $f(x) = \dfrac{6x^4 + x^3 + 1}{-5x + 2x^2}$ is $y = 3x^2$.

Enter Y₁ with care. Use parentheses to enclose both the numerator and denominator, and be careful to use negation, not subtraction. Use thick style for Y₂=3X². Graph with ZDecimal and TRACE. We see a poor picture for the long-run. We need to expand the window. *Note:* The Y₁ equation at the top of the screen is too long to be completely shown.

In this window we see how the long-run values at either side of the graph are similar. We ignore the spike in Y₁ because it is short-run behavior. The settings for this window were $-20 \le x \le 20$ and $-200 \le y \le 1000$.

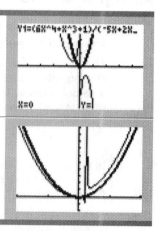

Tip: In defining a rational function, be sure to include parentheses. Use the form Y =(...)/(...) even when it is not necessary.

[398] Show that the long-run behavior of $r(x) = \dfrac{x+3}{x+2}$ is $y = 1$. (This could also be stated as: 1 is a horizontal asymptote.)

Enter the function and use `ZStandard` to see that the function values approach 1 as $x \to \pm\infty$. We again ignore the spike in Y₃ because it is short-run behavior.

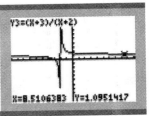

Tip: For Hollywood-style functions that have all their zeros and asymptotes close to the origin, the `ZStandard` window often allows a good view of both short- and long-run behavior.

[400]   Show that the long-run behavior of $r(x) = \dfrac{1}{x^2 + 3}$ is asymptotic to $y = 0$. We also could have said that the behavior of r is like that of $y = 1/x^2$ (which has a horizontal asymptote at zero).

Enter the function and use `ZStandard`. A better window is needed for the short-run behavior, but it does appear that the function values approach 0 as $x \to \pm\infty$. (You may need to trace left and right to convince yourself.)

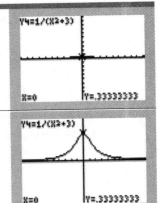

Changing the y settings to $-0.5 \leq y \leq 0.5$ gives a better view of the function values as $x \to \pm\infty$. They approach 0 as a horizontal asymptote. In contrast to the other examples of this section, there is no vertical asymptote. Only rational functions with a denominator that is zero for some x can possibly have a vertical asymptote.

9.5 Short-Run Behavior of Rational Functions

The rational functions are often the hardest graphs to get a good view of because you want to see both the long- and short-run behavior. By finding the zeros of the numerator, you have all the possible zeros of the rational function. By finding all the zeros of the denominator, you have all the possible vertical asymptotes. Combine this with the method of finding long-run behavior from the last section and you should have a good idea of how the graph of a given rational function should look.

The Graph of a Rational Function

Describe the graph of rational function

$$g(x) = \frac{3x + 2}{(x - 1)(x + 2)}.$$

Enter the function and view in a `ZStandard` window. We see the overall pattern we expected: one zero at -2/3, two vertical asymptotes at $x = -2$ and $x = 1$, and a horizontal asymptote at $y = 0$.

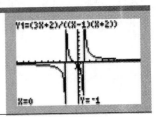

A purist would be upset by the connection of the graph across the two vertical asymptotes. This can be avoided by changing `Connected` to `Dot` in the `MODE` menu. However, as you can see, the graph is less suggestive of the continuity that does exist near the two vertical asymptotes.

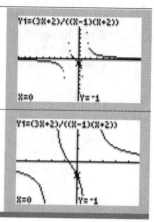

Another means of avoiding the dishonest connection but staying in the `Connected` mode is to be sure that the function is evaluated at undefined points. In this `ZDecimal` window (in `Connected` mode), the connections did not take place because both $x = -2$ and $x = 1$ were evaluated as undefined points. Graph connections are not made across undefined points.

By now you should be sophisticated enough to know that the connection lines across undefined values of a graph should not really be there. It is easier to see them for what they are and to ignore them than going out of your way to change modes or finding special viewing windows to avoid them.

[404]   Describe the graph of rational function

$$r(x) = \frac{x^2 + 2x - 3}{x^2}.$$

Enter the function and view in a `ZStandard` window. We see the overall pattern: one zero at -3, the other at 1. The function is undefined at $x = 0$ and has a vertical asymptotes there. There is a horizontal asymptote at $y = 1$.

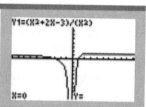

View the same function again in a `ZDecimal` window. The horizontal asymptote appears to be greater than 1 in this window. We must exercise care when finding asymptotes from graphs.

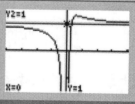

We added the horizontal line `Y₂=1` and use a better window of $-20 \leq x \leq 20$ and $-1.5 \leq y \leq 1.5$. Now we see the behavior of the graph more clearly. Near the origin it crosses the asymptote and then slowly gets close to $y = 1$ from above. Where as for negative x it approaches 1 from below.

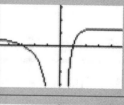

When Numerator and Denominator Have the Same Zeros

[406]   Describe the graph of rational function

$$h(x) = \frac{x^2 + x - 2}{(x - 1)}.$$

We first graph this function in a `ZStandard` window. The zeros of the numerator are $x = -2$ and $x = 1$, but we see graphically that there is no intercept at $x = 1$. The denominator has a zero at $x = 1$, but there is no vertical asymptote. Thus, there is no zero at $x = 1$ because it is undefined.

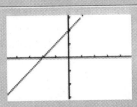

Using a `ZDecimal` window, we see that there is a pixel missing in the graph at $x = 1$. Trace to confirm this. In the `ZStandard` window we could not trace to the exact point $x = 1$, and so there was no missing pixel to warn us about the undefined point.

Tip: When copying a graph such as the one above, you should draw an open circle at the undefined point. This was previously discussed at the end of Section 2.3.

9.6 Comparing Power, Exponential and Log Functions

In this section, we graph and compare three different families of functions. The definition and basic graphing are straightforward. However, you may want to review some techniques that allow you to find good viewing windows. In some cases with multiple functions, there is a good viewing window for one but not all of the functions, and in the end there may not be one window that meets the needs of all the functions in a particular graph.

[412]   Compare $f(x) = x^4$ and $g(x) = 2^x$.

We graph the functions in a `ZDecimal` window. Looking at the screen we see that the power function has larger values except near the origin. We know $g(x)$ has a zero horizontal asymptote as $x \to -\infty$, so $f(x)$ will obviously dominate for negative x. It is unclear which function might dominate as $x \to \infty$.

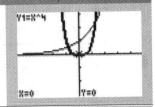

To find the long-run behavior, we look at table values to give us a clue. At $x = 10$, we see that Y_2 lags Y_1, but by $x = 20$, Y_2 is near a million and seems to dominate Y_1 from then on.

(Recall that in E–notation, `1.05E6` = 1,050,000.)

To be sure about the end behavior for positive x, we continue for even larger x-values. Yes, we see that Y_2 continues to dominate, but then we encounter **ERROR**. This is because the Y_2 function values exceeded the limit of the calculator. (No exponent can exceed two digits.)

To set a nice window, we use the million value obtained from our table with $x = 20$. The `Ymin` is set to allow room at the bottom of the screen for trace values.

```
WINDOW
 Xmin=0
 Xmax=20
 Xscl=10
 Ymin=-100000
 Ymax=1000000
 Yscl=0
 Xres=1
```

After graphing, it may not be clear which function is which, so we use `TRACE` and arrow over to find that the power function is dominated by the exponential function. This is a general result: Any exponential function will eventually dominate any power function.

```
Y2=2^X
```
```
X=18.510638    Y=373471.42
```

This final screen, which is essentially the one above, could have been more quickly by setting $0 \le x \le 20$ and using `Zoomfit`. It avoids looking at table values to decide good y window settings.

9.7 Fitting Exponentials and Polynomials to Data

In Section 1.6 we saw how to fit a linear regression equation to data. In this section we show how to fit either an exponential regression equation or a power regression equation to data.

The Spread of AIDS

[416] The table below gives the total number of deaths in the United States from AIDS since 1981. We use $t = 1$ for 1981.

t	1	2	3	4	5	6	7	8
N	159	622	2130	5635	12607	24717	41129	62248

t	9	10	11	12	13	14	15	16
N	90039	121577	158193	199287	243923	292586	340957	375904

One approach to fitting a curve to this data is to use the log function to *linearize* the data. This yields a linear regression equation for the converted data that can then be used to algebraically derive the exponential or power regression formula. This approach is good for conceptual understanding, and your instructor may want you to do the work that way. However, the TI can derive both exponential and power regression formulas in a simple sequence of steps.

Exponential Regression

Enter the data using **STAT Edit**. The data entry in **L1** is not time-consuming but using **seq(** is faster. See Section 6.2.

Use **2nd STAT PLOT** and **ZoomStat** to see a data plot. It is clear that a linear regression equation does not seem appropriate.

On a fresh home screen, use **STAT CALC** and select **ExpReg** (Exponential Regression) to begin the command.

After pasting **ExpReg**, specify the two list names and a **Y=** equation for the regression formula to be pasted into.

The parameters a and b are calculated for the equation **y=a*b^x**. The r and r^2 are measures of correlation as previously described.

Optional step: We check that **Y1** has been properly pasted and that both **Y1** and **Plot1** are selected to be graphed.

This graph shows the data points and the exponential regression curve. The approximation starts out well, but then is a little low in the middle and appears to be too high for larger values. We now try to do better with a power regression equation.

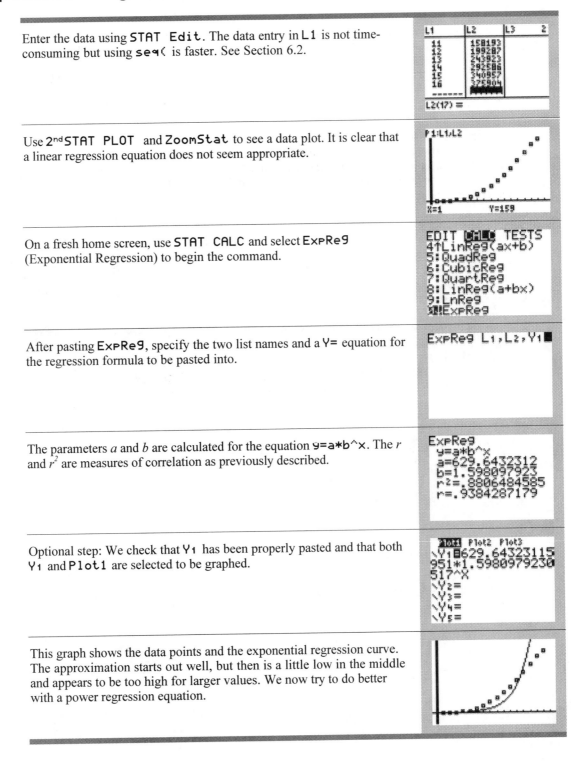

Power Regression

Now select `PwrReg` from the 2ⁿᵈ `STAT CALC` menu and create the command: `PwrReg L1,L2,Y1`.

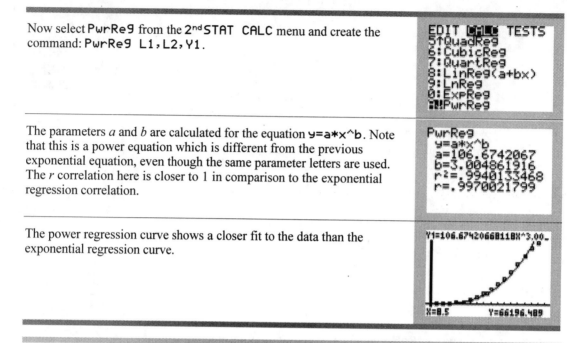

The parameters a and b are calculated for the equation $y=a*x^b$. Note that this is a power equation which is different from the previous exponential equation, even though the same parameter letters are used. The r correlation here is closer to 1 in comparison to the exponential regression correlation.

The power regression curve shows a closer fit to the data than the exponential regression curve.

Tip: There are many other types of regression equations in the `STAT CALC` menu. For a complete description, see Chapter 12 in the *TI–Guidebook*.

CHAPTER TEN

VECTORS AND MATRICES

10.1 Vectors

The TI-83 does not support vector definitions or vector operations, but by thinking of a vector as a list of components, we can use the list features of the TI-83 to work with vectors. Both the TI-89 and the TI-92 have a vector data type and allow some vector operations.

Addition of Vectors

[438] A vector can represent strength and direction of a gravitational force. A spacecraft, thought of as situated at the origin, may experience two vectors of force. One force is eight units along the x-axis, and the other is three units at an angle of -70° with the x-axis. Find the magnitude and direction of the sum of the two force vectors.

Moving the tail of the second vector to the head of the first vector gives a triangle with adjacent sides of 3 and 8 and an included angle of 110° (110° = 180° – 70°). We use the Law of Cosines and Law of Sines to find our results.

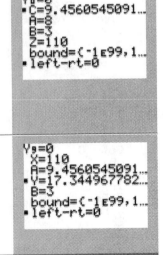

Recall that we have the Law of Cosines stored in Y□, so we use Solver (or solve) to find the length of the third side, which in this case is the magnitude. The steps, which are not shown, are:

1. Set mode to Degree
2. Paste Y□ as the equation
3. Use down-arrow to enter the values of A, B, and Z
4. Arrow up to C
5. Press ALPHA SOLVE.

The magnitude of the sum is C.

To find the angle of declination for the sum vector, change the equation to the Law of Sines (Y₉) and down-arrow to enter the values of X, A, and B. You can see that A has full accuracy, which certainly was not entered by hand. By entering A=C, the value was pasted from the previous result. As usual, the angle Y is found by having the cursor on its line and pressing ALPHA SOLVE.

The above approach is overly complicated and will be simplified in Section 10.3.

10.2 Components of a Vector

There are several ways a vector may be designated. A so-called free vector is given by a pair of rectangular coordinates for its head and another pair of coordinates for its tail. The components of a vector are the coordinate pair given as if the vector had moved its tail to the origin. The polar coordinates of a point also describe a vector whose tail is at the origin.

To facilitate finding components and working with vector operations, we will create a program to use. To enter the program requires extensive pasting and entering of **ALPHA** characters. It is easier to use *TI-Connect™* and enter the program from the keyboard of a computer. It is even easier to have a friend or the instructor let you copy the program using the 2nd**LINK** command. (See the Appendix for details.) The creation and naming of a program were introduced in Section 2.6.

After a program has been named and entry begun, you can exit at any time with 2nd**QUIT**. To check the program code or continue editing, press **PRGM EDIT** and select the desired number.	```EXEC EDIT NEW 1:QF 2▪VECTORS```
This screen contains the first few lines of the program so that you can check your work for proper entry. Long lines are wrapped for a full view. New lines start with a colon.	```PROGRAM:VECTORS :Menu("VECTORS", "POSITION",A,"R, θ",B,"I,J",C,"EX IT",E :Lbl A :Input "X1? ",X :Input "Y1? ",Y```

Vector Program

```
Menu("VECTORS","POSITION",A,"R,θ",B,"I,J",C,"EXIT",E
Lbl A
Input "X1? ",X
Input "Y1? ",Y
Input "X2? ",Z
Input "Y2? ",W
Z-X→I
W-Y→J
R▶Pr(I,J)→R
R▶Pθ(I,J)→θ
Goto D
Lbl B
Input "R? ",R
Input "θ? ",θ
P▶Rx(R,θ)→I
P▶Ry(R,θ)→J
Goto D
Lbl C
Input "I? ",I
Input "J? ",J
R▶Pr(I,J)→R
R▶Pθ(I,J)→θ
Lbl D
ClrHome
Disp "R,θ",R,θ
Disp "I,J",I,J
Lbl E
Stop
```

Help to find commands:

"	ALPHA +
θ	ALPHA 3
ClrHome	PRGM I/O 8
Disp	PRGM I/O 3
Goto	PRGM CTL 0
Input	PRGM I/O 1
Lbl	PRGM CTL 9
Menu(	PRGM CTL C
P▶Rx	2ndANGLE 7
R▶Pr	2ndANGLE 5
Space ␣	ALPHA 0
Stop	PRGM CTL F

The basic idea of this program is to show a menu of three ways to represent a vector. We enter known coordinates in one of the three representation and the vector's coordinates are shown in two forms. Input choices are by **POSITION**, meaning a free vector; by **R,θ**, meaning polar coordinates; or by **I,J**, meaning rectangular components. The output screen shows both the **R,θ** and the **I,J** coordinate representation.

> **Tip:**   If you use the *TI-Connect*™ to enter a program from your computer, be careful about adding spaces at the end of command lines. After the program is transferred to your calculator, extra (unseen) spaces cause syntax errors.

Changing from Polar to Rectangular Coordinates

[444]   A ship travels 200 miles supposedly going due east, but in fact it has traveled 17.4° to the north of east. How far north is the ship from its intended course?

We can enter the vector as R and θ and use the program conversion to find J, the north component.

To start the program, press `PRGM EXEC` and select the program.	`EXEC EDIT NEW`
	`1:QP`
	`2:VECTORS`

The program name is pasted to the home screen. Press `ENTER`.	`prgmVECTORS`

The program menu is shown; select `R,θ`.	`VECTORS`
	`1:POSITION`
	`2:R,θ`
	`3:I,J`
	`4:EXIT`

Enter the R and θ as shown. If you are not in `Degree` mode, then you should enter θ=`17.4°`. Even when you are entering data to a program, you can use 2ⁿᵈ`ANGLE` to paste the degree symbol.	`prgmVECTORS`
	`R? 200`
	`θ? 17.4▪`

The answer is J ≈ 59.8. The values shown for R, θ, I, and J are stored in memory, so if needed, they can be used for further calculations.	`R,θ`
	`            200`
	`           17.4`
	`I,J`
	`    190.8480657`
	`    59.80815845`
	`          Done`

> **Tip:**   After you see the program output on the home screen, you can press `ENTER` and the program starts again. This is very handy for dealing with several vectors.

Finding Vector Components

[447]   Find components of a vector whose tail is (-3,1) and whose head is (2,4).

By pressing `ENTER` from the previous screen, the program will run again. This time select `POSITION`.

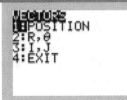

Enter the pairs of coordinates. (Notice that the top of the screen still has the leftover values showing from the last use. The program could clear the input screen by inserting **ClrHome** as the first line of the program. However, it is often convenient to see previous values, so this has not been done.

The vector components are (5, 3), so the vector could be written as $5\vec{i} + 3\vec{j}$ In this case the **R,θ** are not needed, but the program always shows both forms.

10.3 Applications of Vectors

So far the vectors have been only two-dimensional and handled by using a variable for each dimension, as seen by our use of **I** and **J**. For higher dimensional vectors, we can use lists.

Vector Arithmetic

[451]   Suppose $\vec{R} = (domestic, international, videotape)$ is a revenue vector, in millions of dollars, for a movie. The movies *Shrek*, *Pearl Harbor*, and *The Mummy Returns* had vectors:

$$\vec{s} = (268, 198, 436), \quad \vec{t} = (199, 252, 301), \text{ and } \vec{u} = (202, 227, 192).$$

Find the total revenue vector for all three of these movies.

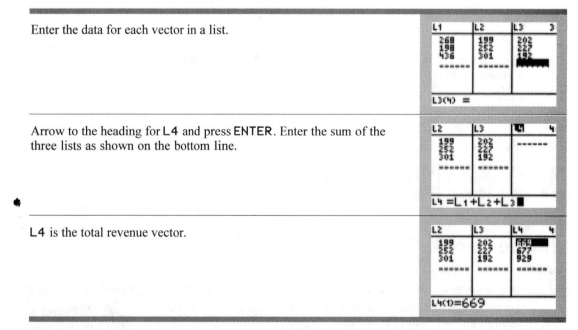

Enter the data for each vector in a list.

Arrow to the heading for **L4** and press **ENTER**. Enter the sum of the three lists as shown on the bottom line.

L4 is the total revenue vector.

[453]   In computer graphics, a pixel has an (*x*, *y*) location and is often thought of as being moved by a given distance in a certain direction. This movement is described by a vector (*r*, *θ*). For example, suppose the point (3, 5) is moved three units in the 70° direction. (The angle is measured counterclockwise from the *x*–axis.)

We have used the VECTORS program to enter the movement vector R=3 and θ=70 (De9ree mode). This gives us the I,J components of the movement which we can add to the point.	R,θ 3 70 I,J 1.02606043 2.819077862 Done
Call the program again by pressing ENTER and select I,J.	VECTORS 1:POSITION 2:R,θ 3:I,J 4:EXIT
We reuse the full accuracy of the previous I and J by entering them as names when they are summed with the original point.	3 70 I,J 1.02606043 2.819077862 Done I? I+3 J? J+5■
This output screen shows the new position of the point as I,J.	R,θ 8.794722349 62.75596142 I,J 4.02606043 7.819077862 Done

[438] In a similar, way we can redo the rocket vector sum example from Section 10.1. The new approach is to just add the components of the two vectors to find the resultant vector. We avoid having to think about moving vectors and forming triangles and using both the Law of Sines and the Law of Cosines. This approach is rocket science made simple.

We use the VECTORS program (and De9ree mode) to enter R=3 and θ=70. (We think of declination as positive and avoid negative signs.)	R,θ 3 70 I,J 1.02606043 2.819077862 Done
We add the horizontal vector (I=8 and J=0) to the previous vector, whose coordinates are still stored in I and J.	R,θ 3 70 I,J 1.02606043 2.819077862 Done I? I+8 J? J■
We read the sum using the R, θ vector. This corresponds to the result found in Section 10.1. We have a force of R ≈ 9.456 acting at an angle of declination θ ≈ 17.345.	R,θ 9.456054509 17.34496778 I,J 9.02606043 2.819077862 Done

10.4 The Dot Product

Definition: If $\vec{u} = (u_1, u_2, \cdots, u_n)$ and $\vec{v} = (v_1, v_2, \cdots, v_n)$ are *n*-dimensional vectors, then the *dot product* is defined as

$$\vec{u} \cdot \vec{v} = u_1 v_1 + u_2 v_2 + \cdots + u_n v_n.$$

[456] A car dealer sells five models of a car. The number of cars sold in a week is given by the vector $\vec{C} = (22, 14, 8, 12, 19)$, and the price of each model is given by the vector $\vec{P} = (14, 18, 35, 42, 27)$. Find this week's revenue.

On the home screen, enter the vectors as lists. (You could also use STAT Edit.)	{22,14,8,12,19}→L₁ {22 14 8 12 19} {14,18,35,42,27}→L₂ {14 18 35 42 27}
From the 2ⁿᵈLIST MATH menu select sum(.	NAMES OPS **MATH** 1:min(2:max(3:mean(4:median(5▓sum(6:prod(7↓stdDev(
Complete the sum command as shown with a multiplication of the two lists. Multiplication of lists is term by term, so by summing the multiplication list you have the dot product.	L₁ {22 14 8 12 19} {14,18,35,42,27} →L₂ {14 18 35 42 27} sum(L₁*L₂) 　　　　1857

[458] Find $\vec{v} \cdot \vec{w}$ where $\vec{v} = 3\vec{i} + 4\vec{j}$ and $\vec{w} = 2\vec{i} + 5\vec{j}$. Find $\| \vec{v} \| = \sqrt{\vec{v} \cdot \vec{v}}$ and $\| \vec{w} \| = \sqrt{\vec{w} \cdot \vec{w}}$.

We find the dot product using lists. However, it is hardly worth putting two-dimensional vectors into lists. In this simplistic case, you could enter the calculation directly from the dot product definition, 3*2+4*5 = 26.	{3,4}→L₁ 　　　　{3 4} {2,5}→L₂ 　　　　{2 5} sum(L₁*L₂) 　　　　26
We can also find the norm (or length) of a vector by using the formula; this applies no matter what the dimension of the vector.	√(sum(L₁*L₁)) 　　　　5 √(sum(L₂*L₂)) 　　5.385164807
In the two-dimensional vector case, the norm is just the **R** of the polar coordinate. To show this, we use the **VECTORS** program. Enter **I=2** and **J=5** for the second vector to see that **R=5.385**... is the norm of the vector.	R,θ 　　5.385164807 　　68.19859051 I,J 　　　　2 　　　　5 　　　Done

> *Tip:* Values shown on the VECTORS program screen are stored in variable names and remain until the next calculation of the program or changes are made to that variable. A value from a home screen calculation (in Ans) remains only until the next calculation.

10.5 Matrices

> *Tip:* The matrix menu key on the old TI-83 has been replace by the APPS key on the TI-83 Plus. TI-83 non-Plus users should translate references to 2ndMATRIX to be MATRX. (There is no I in MATRX on the non-Plus TI-83.)

Matrix Arithmetic

A matrix is a rectangular array of numbers. It can, like vectors, be multiplied by a scalar. Matrices of the same dimensions can be added and subtracted.

Addition and scalar multiplication

[469]   Evaluate the expressions (a) $5A$ (b) $-2B$ (c) $A+B$ (d) $B-3A$, given that:

$$A = \begin{pmatrix} 3 & 7 \\ 2 & -1 \end{pmatrix} \text{ and } B = \begin{pmatrix} 1 & -5 \\ 0 & 8 \end{pmatrix}.$$

Press 2ndMATRIX to see the menu of matrix names. Matrix names are enclosed by square brackets. Right-arrow to the EDIT menu as shown here.	NAMES MATH EDIT 1:[A] 2:[B] 3:[C] 4:[D] 5:[E] 6:[F] 7↓[G]
Press ENTER to start the edit. Any matrix not been previously defined will have the dimension 1×1. The dimension of a matrix is given as *row × column*.	MATRIX[A] ×1 [0]
Enter the dimension 2 and press ENTER. Press 2 ENTER again. The cursor is now in the matrix and ready to enter the matrix values. Initially matrix values are set to zero.	MATRIX[A] 2 ×2 [0 0] [0 0] 1,1=0
Matrix values are most easily entered from left to right using the ENTER key after each entry. At the end of a row the cursor proceeds down and to the left of the new row, just like reading a newspaper. To make corrections you can use the arrow keys to position the cursor on any entry. Use 2ndQUIT when the matrix is entered correctly or use 2ndMATRIX EDIT to start editing another matrix.	MATRIX[A] 2 ×2 [3 7] [2 -1] 2,2=-1

Matrix **[B]** is shown fully entered. Now that the matrices are filled, we can do the required matrix arithmetic.

```
MATRIX[B] 2 ×2
[ 1      ‾5      ]
[ 0      8       ]



2,2=8
```

To calculate 5*A*, start with a clear home screen and enter the scalar 5. (Screen not shown.) Paste **[A]** from the **2ⁿᵈMATRIX NAMES** menu. Unfortunately there are no handy matrix names listed anywhere on the keyboard keys like there are for list names.

```
NAMES MATH EDIT
1:[A]      2×2
2:[B]      2×2
3:[C]
4:[D]
5:[E]
6:[F]
7↓[G]
```

When the calculation is ready press **ENTER**; the matrix answer is shown. Notice that the matrix format starts with **[[** and ends with **]]** and that there is a row on each line.

```
5[A]
          [[15 35],
           [10 ‾5]]
-2[B]
          [[‾2 10 ],
           [0  ‾16]]
■
```

For calculating part (c) and (d) repeat the above procedure. For each matrix name you must return to the **2ⁿᵈMATRIX** menu.

```
[A]+[B]
          [[4 2],
           [2 7]]
[B]-3[A]
          [[‾8 ‾26],
           [‾6 11 ]]
■
```

> **Tip:** When you enter the **2ⁿᵈMATRIX** menu you are in the **NAMES** menu. To edit **[B]** you must arrow to **EDIT** before making the **[B]** matrix choice. If you first arrow down to **[B]** and then arrow to **EDIT**, the matrix selection resets to **[A]** and you have to arrow down to **[B]** again.

Matrix multiplication

[468]   Computer graphics often require the rotation of an object about a fixed point. As an example we show how a pixel, thought of as a position vector, can be rotated about the origin by a given angle. If the pixel coordinates are (2, 3) and the rotation is 30 degrees, then the new pixel coordinates will be given by a matrix multiplication.

$$R\vec{P}_0 = \begin{pmatrix} \cos\theta & -\sin\theta \\ \sin\theta & \cos\theta \end{pmatrix} \cdot \begin{bmatrix} x_0 \\ y_0 \end{bmatrix}$$

Enter the rotation matrix **[A]** as a 2×2 matrix. The first entry was **cos(30°)** but it was automatically calculated to its decimal value. The second value being entered will also be immediately calculated.

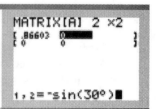

```
MATRIX[A] 2 ×2
[ .86603  0      ]
[ 0       0      ]



1,2=‾sin(30°)■
```

Finish entering the rotation matrix [A] as a 2×2 matrix.	MATRIX[A] 2 ×2 [.86603 -.5] [.5] 2,2=.8660254037...
Enter the position vector as a 2×1 matrix [B]. Press 2ⁿᵈQUIT to return to the home screen. If [B] was previously a 2×2 matrix change the dimension to 2×1 on the top line.	MATRIX[B] 2 ×1 [2] [3] 2,1=3
Press 2ⁿᵈMATRIX to see on the NAMES menu that you correctly defined the two matrix dimensions. Use this screen menu to paste matrix names for matrix calculations.	NAMES MATH EDIT 1:[A] 2×2 2:[B] 2×1 3:[C] 4:[D] 5:[E] 6:[F] 7↓[G]
The position vector rotates from (2, 3) to (0.232, 3.598).	[A][B] [[.2320508076] [3.598076211]]

Solving linear systems of equations

We show how to use matrices to solve a system of simultaneous linear equations. This example is from the Tools section in Chapter 1 of the text. Commonly, a system of linear equations has n equations and n unknown variables. We will show an example for $n = 2$, but the same technique can be used for larger values of n.

[55]   Solve the system $\begin{cases} y + \dfrac{x}{2} = 3 \\ 2(x + y) = 1 - y \end{cases}$

First rewrite the two equations so that the unknowns are all on the left and the constant terms are all on the right,

$$\begin{cases} \dfrac{1}{2}x + y = 3 \\ 2x + 3y = 1 \end{cases}$$

The coefficients of the unknowns are written as one matrix, A, and the constants are written as a second matrix, B. The simultaneous solution of the two equations is found by the simple matrix arithmetic $A^{-1}*B$.

Enter the equation's coefficients in [A].

```
MATRIX[A] 2 ×2
[ 5          ]
[ 2          ]

2,2=3
```

Repeat the matrix definition process to enter the 2×1 matrix [B]. This matrix is a column matrix with the constants from the right-hand side of the system of equations.

```
MATRIX[B] 2 ×1
[          ]

2,1=1
```

Press 2ⁿᵈMATRIX and to see that you have defined the two matrices because they now show their dimensions on the NAMES menu. Use this screen menu to paste matrix names, like [A], onto the home screen for matrix calculations.

```
NAMES MATH EDIT
1:[A]    2×2
2:[B]    2×1
3:[C]
4:[D]
5:[E]
6:[F]
7↓[G]
```

Enter the calculation shown. Both [A] and [B] must be pasted from the 2ⁿᵈMATRIX NAMES menu, and $^{-1}$ is pasted by using the x^{-1} key. Read the answer from the matrix shown: $x = -16$ and $y = 11$.

```
[A]⁻¹[B]
        [[-16]
         [11 ]]
```

Tip: Do not enter a matrix name on the keypad. If you type 2ⁿᵈ[ALPHA A 2ⁿᵈ], the screen will show [A] but it will cause a syntax error. Use 2ⁿᵈMATRIX NAMES and paste a matrix name.

Tip: It may have occurred to you that an $n \times 1$ or $1 \times n$ matrix could be used to represent vectors. Matrices that are just one column or just one row are often referred to as vectors.

CHAPTER ELEVEN

SEQUENCES AND SERIES

11.1 Sequences

A sequence is a list of numbers. In an arithmetic sequence, the terms increase (or decrease) by adding a fixed amount. In a geometric sequence, each term is a constant multiple of the preceding term.

Arithmetic sequence

[477] Write a formula for the sequence of annual odometer readings for a car that is driven 8000 miles per year and had gone 15,000 miles when bought. Show the first seven years.

In Section 6.2, we introduced the `seq(` command in the 2nd`LIST OPS` menu. We use this command again to generate a sequence.
 	
    `seq(` *expression, variable, begin, end, increment)* .
 	
Since sequence terms are numbered 1, 2, 3... the *increment* is 1 and can be omitted.
	`NAMES OPS MATH`
	`1:SortA(`
	`2:SortD(`
	`3:dim(`
	`4:Fill(`
	`5:seq(`
	`6:cumSum(`
	`7↓ΔList(`

Complete the command as shown to create a sequence of odometer readings. Since seven years from the date of purchase is the start of the eighth year we use 8 for the *end* value. In general using X as the variable saves keystrokes, but we used N out of respect for tradition.	`seq(15000+(N-1)*`
	`8000,N,1,8)`
	`{15000 23000 31...`

Use the right arrow to scroll the list and find the final mileage is 71,000 at the start of year eight.	`seq(15000+(N-1)*`
	`8000,N,1,8)`
	`.00 63000 71000}`

Geometric sequence

[479] Write a formula for a salary sequence that starts at $40,000 and increases by 5% each year. Show your salary after ten years.

Since we are dealing with dollars, we change decimal rounding to 2 in the MODE.

Like the arithmetic sequence we use the seq(command to create the geometric sequence. Since after ten years from hire is the start of the eleventh year we use 11 for the *end* value.

Use the right arrow to scroll the list and find the salary is $65,155.79 at the start of year eleven.

Other ways to create sequences

Using a function and table

Use the Y= screen to define the continuous function version of the geometric sequence.

Use 2nd TBLSET to start the table at 1 and increment by 1.

Press 2nd TABLE to see the table. Use the arrow keys to find the desired value.

Tip: The function value on the bottom line is always shown in full decimal accuracy, regardless of the MODE decimal setting.

Using the sequence function mode

The arithmetic sequence values can be found by using the Seq option in the MODE menu.

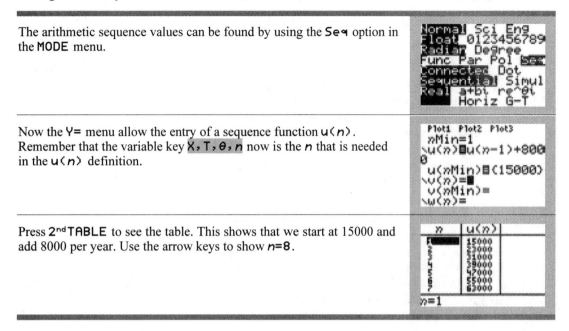

Now the Y= menu allow the entry of a sequence function u(n).
Remember that the variable key X,T,θ,n now is the n that is needed in the u(n) definition.

Press 2nd TABLE to see the table. This shows that we start at 15000 and add 8000 per year. Use the arrow keys to show n=8.

11.2 Arithmetic Series

[482] Lefkos the landscaper stacks timbers to make an upward-sloping terrace. He starts with one timber, then adds a stack of two, then a stack of three, and so on to build the terrace. How many timbers does he need if there are six stacks in the terrace?

We complete the seq(command as shown to create a sequence of the number of timbers in each stack.

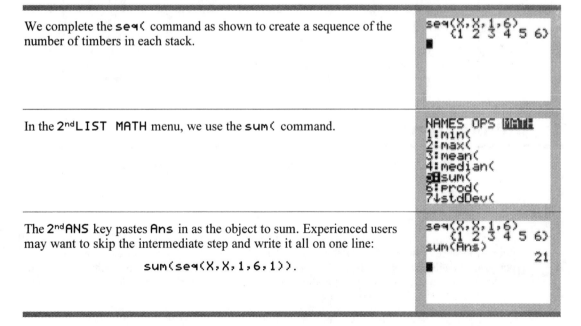

In the 2nd LIST MATH menu, we use the sum(command.

The 2nd ANS key pastes Ans in as the object to sum. Experienced users may want to skip the intermediate step and write it all on one line:

$$sum(seq(X,X,1,6,1)).$$

Sum Notation

It is useful to be able to translate back and forth between the **seq** command and sigma notation. The previous **seq(X,X,1,6)** command has sigma notation: $\displaystyle\sum_{x=1}^{6} x$.

[485] $\displaystyle\sum_{i=1}^{20} (2i - 1)$ is the sigma notation for the first 20 positive odd numbers. Find the sum.

The translation from the sigma notation to the **seq** command is shown. The increment is not specified in sigma notation; but it is always 1.

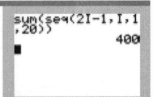

Tip: Any letter can be used for the variable. In the previous example, we followed the notation of the stated problem, but it required that we use the **ALPHA** key. This extra keystroke can be avoided by using the variable **X**, as much as possible.

Falling Objects

[485]   An application of arithmetic sequences is finding the distance traveled by a falling object. The distance traveled per second by a falling object is given by the sequence {16, 48, 80, ...}. The linear function $y = 32x - 16$ can generate these values. Find the total distance fallen for each second.

First, we enter the sequence to check its correctness. Second, we use the **cumSum** command in the **CATALOG** (or **2ⁿᵈ LIST OPS** menu) to generate the cumulative sums.

11.3 Finite Geometric Series

A finite geometric series is a sum of the form: $S_n = a + ax + ax^2 + \cdots + ax^{n-1}$.

IRAs

[489]   A common application is an Individual Retirement Annuity (IRA). The balance of an IRA is a geometric series where a is the annual deposit and x is the growth rate. For example, depositing $2000 per year and getting a 6% return, we can find the balance in 5 years, right after a payment has been made.

$$Q_{10} = 2000 + 2000(1.06) + 2000(1.06)^2 + \cdots + 2000(1.06)^5 = \sum_{n=0}^{5} 2000(1.06)^n .$$

Once a geometric series is translated to sigma notation, the sum is easily found. The time spent changing to sigma notation is usually faster than explicitly entering the term-by-term sum. We find the sequence using seq(in the 2ndLIST OPS menu then we sum the sequence using sum(in the or 2ndLIST MATH menu.

A more efficient approach is to find the sum using a formula. A geometric series can be calculated directly from the formula:

$$S_n = a + ar + ar^2 + \cdots + ar^{n-1} = \frac{a(1-r^n)}{1-r}.$$

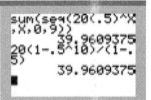

Drug levels

[491]   A patient is given a 20-mg injection. Each day the patient's body metabolizes 50% of the drug present. Write a geometric series that gives the drug level in the body after the tenth injection.
The geometric series is

$$Q_{10} = 20 + 20\left(\tfrac{1}{2}\right) + 20\left(\tfrac{1}{2}\right)^2 + \cdots + 20\left(\tfrac{1}{2}\right)^9 = \sum_{x=0}^{9} 20\left(\tfrac{1}{2}\right)^x.$$

We find the sum in two ways. First with sum(seq(...)) and then directly using the sum formula.

Tip: While the nested sum(seq(...)) command is efficient, it does not show the sequence that is being summed. It is prudent to generate the list first, check the list for correctness, and then sum the list using sum(Ans).

11.4 Infinite Geometric Series

We just found the finite sum of a geometric series by means of summing the terms and then by using a formula. For an infinite geometric series, summing all the terms is impossible and we must use a formula.

[493]   We concluded the previous section with the example of a patient given a daily 20-mg injection and metabolizing 50% of the drug present. Write a geometric series that gives the drug level in the body after many injections.

As the treatment goes on for more and more days, the level gets closer and closer to 40. After 100 injections, the sum is so close to 40 that it exceeds the calculator accuracy and it rounds up to 40.

Tip: The last calculation takes several seconds during which time a busy indicator will scroll in the upper right corner of the screen. If a command such as sum(seq(...)) is taking too long to calculate, you can press **ON** to break.

The infinite sum of the geometric series is calculated directly from the formula

$$S = \sum_{n=1}^{\infty} ax^n = \frac{a}{1-x} = \frac{20}{1-(0.5)} = 40.$$

Using the sum formula in this case is clearly more efficient.

The Present Value

[496] Alex Rodriguez signed a contract with the Texas Rangers that had a signing bonus of $2 million per year for five years. The present value (using a 5% return rate) of his contract, is calculated as a geometric series:

$$Q_5 = 2 + 2\left(\tfrac{1}{1.05}\right) + 2\left(\tfrac{1}{1.05}\right)^2 + 2\left(\tfrac{1}{1.05}\right)^3 + 2\left(\tfrac{1}{1.05}\right)^4 = \sum_{n=0}^{4} 2\left(\tfrac{1}{1.05}\right)^n.$$

Since the sum has only 5 terms we could directly enter it. But since we have the sigma-notation formula we use that to find the sum. Either way we see that about $9 million is enough for the team owner to deposit to be able to meet all 5 payments.

The present value of paying Rodriguez $2 million *forever*, can be algebraically derived as $S = 2/(1 - 1/1.05) = 42$. This value is not so easy to derive by using a large sum like we did in the previous drug example. In this case, the sum of 100 terms takes six times longer and is still not very close to the limit value.

Tip: Don't count on finding an infinite sum by choosing a large end value in a sum(seq(...)) command.

CHAPTER TWELVE

PARAMETRIC EQUATIONS AND CONIC SECTIONS

12.1 Parametric Equations

We have seen that the MODE menu lists the function options Func Par Pol Seq, and we have used all but Par so far. Par is the parametric function option. The parametric way of defining a function is by using a variable t and defining the x-y position in terms of t. Thus, the Y= menu changes to define function pairs like X_{1T} and Y_{1T}, the variable key X,T,θ,n pastes the variable T, and the WINDOW screen begins with T-settings.

Parametric Graphing

[506]   Graph the path of a robot that begins at the point (1,0) and follows a path given by the equations:

$$x = \cos t, \text{ and } y = \sin t, \text{ where } t \text{ is in minutes, } 0 \le t \le 6.$$

We start in the Par mode. We will then take the standard approach to graph a function: define the function, set the window, and graph.	Normal Sci Eng Float 0123456789 Radian Degree Func Par Pol Seq
We define a single graph using a pair of functions, one for x and one for y. The variable in parametric mode is T.	Plot1 Plot2 Plot3 \X₁ᴛ■cos(T) Y₁ᴛ■sin(T) \X₂ᴛ=
We use ZDecimal to set the viewing window. However, the problem stipulates $0 \le t \le 6$. We reset T values after we are satisfied with the X and Y window settings.	WINDOW Tmin=0 Tmax=6■ Tstep=.1 Xmin=-4.7 Xmax=4.7 Xscl=1 ↓Ymin=-3.1
Trace the path to the end. It is circular but, because the t interval does not extend to 2π, we see that the robot stopped just short of making it back to the start.	X₁ᴛ=cos(T) Y₁ᴛ=sin(T) T=6 X=.96017029 Y=-.2794155

Tip: The previous example can be altered by changing Tmax to other values. Also, by changing X_{1T}=cos(2T) and Y_{1T}=sin(2T), the robot will stay on the same path but go twice as fast.

Archimedean Spiral Revisited

[509] In Section 7.5 a polar graph of the Archimedean spiral was shown. The graph of this spiral has parametric equations:

$$x = t \cos t, \ y = t \sin t, \text{ where } 0 \leq t \leq 2\pi.$$

After defining the functions, the graph is drawn in a ZStandard window. The ZStandard automatically sets T values as $0 \leq t \leq 2\pi$. The other ZOOM settings do not reset the T values from their current state.

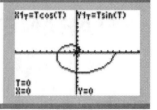

Tip: In parametric mode, the usual graph styles are available, but shading is not allowed.

A Parametric Walk Around the Block

[511] Graph the motion of a particle starting at the origin and "walking" around a unit square in 4 seconds.

For this example, we use the logical expression restrictions introduced in Section 2.3 to make some rather hard to read definitions for our functions. The trick is to define the function values over the four intervals: $0 \leq t < 1$, $1 \leq t < 2$, $2 \leq t < 3$, $3 \leq t < 4$. In the definition shown below, the four intervals are written in TI notation as: (0≤T)(T<1), (1≤T)(T<2), (2≤T)(T<3), (3≤T)(T<4). An alternative TI interval notation is: (0≤T and T<1); this has not been used because it is longer and the definitions would not fit on a single screen.

In this definition, the first term in the tricky sum that defines X_{1T} is T(0≤T)(T<1). This means that the function will be T for $0 \leq T < 1$, but it will be 0 otherwise. The logical expressions control which formula is being used. For the graph style, we specified thick.

After entering the definitions, we use WINDOW to set Tmin=0, Tmax=4, and Tstep=.1. Graph with ZDecimal. Press TRACE to produce the screen shown. Notice that both definitions show across the top, but they are truncated, because they are so long.

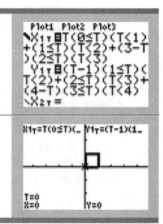

Tip: The screen appearance of a parametric equation name differs slightly depending upon whether it is on the Y= screen as Y_{1T} or the graph screen as $Y1_T$.

Tip: Many beautiful curves can be drawn using the parametric mode. Try the Lissajous figure: $x = \cos(3t)$, $y = \sin(5t)$ with a ZStandard and a ZoomIn.

12.2 Implicit Defined Curves and Circles

Curves created by slicing (sectioning) a cone, are appropriately called conic sections. We have already learned to graph one conic, the parabola. We demonstrate how any function defined in the Func graph mode can be also graphed using the parametric mode (Par). In this section, all angles are measured in radians, so be sure to have the angle mode set to Radian.

Parabola

To graph $y = f(x)$ in the parametric mode, use $x = t$, and $y = f(t)$. For example, the parabola has the equation $y = f(x) = x^2$, so we define $X_{1T}=T$ and $Y_{1T}=T^2$.

After entering the above parametric equations, we set the window as shown. This is accomplished by pressing ZDecimal (which draws a graph we ignore) and then using WINDOW to reset the T values to be the same as the X values.

Pressing TRACE allows us to see the equation definitions. Notice that the trace starts at the Tmin value, which in this case is a point off the screen. (In contrast, TRACE starts at the screen center in the Func mode.)

In Section 8.2, we introduced DrawInv to graph the inverse of a function. In the parametric mode, inverses are simple to graph, and they can even be traced. Simply switch the X and Y definitions. For example switching these equations, as shown at the top of the screen, draws the inverse of the parabola.

Circles

For $r > 0$, the parametric equations of a circle of radius r centered at the point (h, k) are

$$x = h + r \cos t \text{ and } y = k + r \sin t, \ 0 \le t \le 2\pi$$

[515]   Graph the parametric equations $x = 5 + 2\cos t$ and $y = 3 + 2\sin t$.

In Par mode, enter the two equations and graph with ZStandard. This does not look like a circle! The problem is that the ZStandard is not a square setting. It does give us a good idea about the center and what might be needed to have a better window.

For the window shown, we first used ZSquare and then changed Xmin=0 and Ymin=0 to show only the first quadrant.

12.3 Ellipse

The parametric equations of an ellipse centered at (h, k) are:

$$x = h + a\cos t, \quad y = k + b\sin t, \text{ for } 0 \le t \le 2\pi.$$

[518]   Graph the parametric equations $x = 7 + 5\cos t$ and $y = 4 + 2\sin t$.

In this graph we used the window settings from the previous example. These settings are square so that we are assured that this is a noncircular figure.

Tip: As in the parabola example, we can switch the major and minor axes of an ellipse by simply interchanging the equation definitions.

12.4 Hyperbola

The parametric equations of a hyperbola centered at (h, k) are:

$$x = h + a\sec t, \quad y = k + b\tan t, \text{ for } 0 \le t \le 2\pi.$$

[523]   Graph the equation $\dfrac{(x-4)^2}{9} - \dfrac{(y-7)^2}{25} = 1$.

We first translate this equation to parametric form. The center is found from the numerators, $(h, k) = (4, 7)$. The values of a and b are found from the denominators as $a = \sqrt{9} = 3$ and $b = \sqrt{25} = 5$.

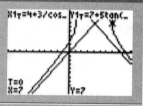

Enter the functions with special care since there is no secant function. $X_{1T} = 4 + 3/\cos(T)$ and $Y_{1T} = 7 + 5\tan(T)$. Use a `ZStandard` window to get a first view.

The window settings for T have been left as $0 \le t \le 2\pi$, but the X and Y window values have been set to give a better view.

```
WINDOW
↑Tstep=.1308996...
 Xmin=-6
 Xmax=14
 Xscl=2
 Ymin=-10
 Ymax=25
 Yscl=5
```

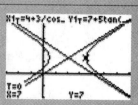

This shows the characteristic figure for a hyperbolic function. You may be wondering about the X-lines in the middle of the screen. This is the connection across discontinuities that we previously encountered for rational functions. As before, use `Dots` mode to make the lines disappear or just think of them as asymptotes.

12.5 Hyperbolic Functions

The hyperbolic sine and hyperbolic cosine are defined by

$$\cosh(x) = \frac{e^x + e^{-x}}{2} \quad \text{and} \quad \sinh(x) = \frac{e^x - e^{-x}}{2}.$$

Graphing the Hyperbolic Functions

[526]   Graph the cosh and sinh functions.

Use the Func graphing mode. Shown here is the cosh(in the CATALOG. Paste this to Y1 to define the function.

We graph in a ZDecimal window.

Similarly, we define Y2=sinh(X) and graph.

The Identity cosh²x - sinh²x = 1

[528]   The identity $\cosh^2 x - \sinh^2 x = 1$ can be graphed like its counterpart $\cos^2 x + \sin^2 x = 1$.

The expression $\cosh^2 x$ can be written in any of the following ways: cosh(X)², (cosh(X))², or as cosh(X)^2. The left side of the identity is written as Y1.

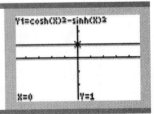

Parameterizing the Hyperbola Using Hyperbolic Functions

[528]   Compare the curve parameterized by the equations

$$x = \cosh(t), \quad y = \sinh(t), \quad \text{for} -\infty < t < \infty$$

and the hyperbola defined by $x^2 - y^2 = 1$.

First graph $x^2 - y^2 = 1$ as the parametric equations $x = \sec t$, $y = \tan t$.

Enter the two trigonometric functions in X_{1T} and Y_{1T}. Use **ZDecimal** and $-\pi \le t \le \pi$. We see the graph is centered at the origin and has two branches opening left and right. Think of the lines through the origin as asymptotes.	

Turn off the first pair. Now enter the second pair of equations as hyperbolic functions in X_{2T} and Y_{2T}. Graph to see that this is the same as the rightward opening branch of the hyperbola graphed above.	

APPENDIX

Internet Addresses

The main Internet address for Texas Instruments calculators is

http://education.ti.com

This home page has a link to the home page of the TI-83 Plus Silver Edition, whose direct address is

http://education.ti.com/us/product/tech/83pse/features/features.html

These addresses are current as of May 2003. Web addresses and designs do change, but normally old addresses are maintained so that these addresses should be active for the life of this manual.

Menu Options at the TI-83 Plus-SE Website

features is the home page of the TI-83 Plus-SE and lists relevant features.

apps is a list of apps (applications) that can be downloaded to the TI-83 Plus. (Not an option for non-Plus) Here is sample of some helpful precalculus apps:
- o Catalog Help - Provides easy access to calculator syntax
- o Area Formulas - Reinforce and review area formulas
- o Conic Graphing - Graph the four conic shapes
- o Guess My Coefficients - Identify coefficients from a graph
- o Inequality Graphing - Graphing inequalities is made simple
- o Polynomial Root Finder and Simultaneous Equation Solver
- o Transformation Graphing - Helps improve graphing comprehension
- o What is My Angle - Improve comprehension of the measure of angles through real-world examples

community center is where you can access to resources and/or join a discussion group.

FAQs directs you to the help resources for specific problems.

guidebooks has links to download missing manuals.

accessories are listed.
- o TI Keyboard – A new and useful item
- o TI-GRAPH LINK™ Cables
- o CBL 2™ and CBR™ data collection devices
- o Color Slide Cases
- o TI ViewScreen™ Panel and TI-Presenter™ video interface

software is available for downloading.
- o TI Connect™ Software
- o TI-GRAPH LINK™ Tools
- o Software Development Kit - program your own Apps
- o Program Archive - download programs created by other users
- o Key Fonts
- o StudyCards™ Stacks
- o Tutorials -interactive tutorials

where to buy and **special offers** – online shopping!

Linking Calculators

The essentials of linking are presented in the *TI–Guidebook* and will not be repeated here. But we include the following tips.

♦   The end-jack must be pushed firmly into the socket. There is a final click you can feel as it makes the proper connection.

♦   If you are experiencing difficulty connecting, turn off both calculators, check the connection, and then turn them on and try again. If available, try other cables or calculators.

♦   When selecting items, the cursor square is hardly visible when the selection arrow is on the same item. As you arrow off of the item, double-check whether or not it has been selected.

♦   If you are required to drain your calculator memory before entering an exam, use 2ⁿᵈLINK Back Up to keep a copy on some other calculator. Even better, store it on your computer, which is the next topic.

Linking to a Computer

The *TI-Graph Link*™ is a cable and *TI-Connect*™ is the software needed to connect your TI calculator to a PC or Macintosh computer. The software is available on the Internet, so you can order only the cable. The cable is included with the TI-83 Plus-SE. There are many advantages to using *TI-Graph Link*™ and *TI-Connect*™.

♦   This is the best way to back up your work.

♦   You can download apps from the TI website.

♦   It is the preferred way to write and edit programs.

♦   You can download and transfer programs from the Internet archives.

♦   It allows you to capture the screen in a form for direct printing or use in a word processor.

Troubleshooting

Nothing shows on the screen

♦   Check the contrast.

♦   Check the ON/OFF button.

♦   Pull out the batteries and correctly reinsert the batteries.

♦   As a last resort, remove the backup battery and reinsert all the batteries. *Warning:* This will erase all memory, including programs.

Nothing shows up on the graph screen except the axes

♦   Press TRACE to see if a function is defined but outside the window.

♦   There may be no functions selected.

♦   The function may be graphed along either axis and need the window reset.

◆ If there is a busy indicator (a running line) in the upper right corner of the screen, then the TI may be still calculating. Press **ON** if you can't wait.

◆ If there is a pause indicator (a twinkling line) in the upper right corner of the screen, then the TI is paused from a program. Press **ENTER** to continue.

◆ If there is a full checkerboard cursor, then you have a full memory. You need to delete something; choose some things you no longer need or copy them to your computer.

Nothing shows up in the table

You may have the **Ask** mode set and need to either enter *x*-values or change to **Auto** in the 2ⁿᵈ**TBLSET**.

Check to see that a function is selected.

I get a syntax error screen

The most common errors are

◆ Parentheses mismatch. Count and match parentheses carefully.

◆ Subtraction versus negative symbol. For example, the subtraction sign cannot be used to enter **Xmin=-10**.

◆ Pasting a command in the wrong place. For example, a program name must be on a fresh line.

I get an error message

This can cover the widest array of problems. Read the message carefully; it will tip you off to the kind of error you are looking for. If you have no idea what could have caused it, consult the appendix of the *TI–Guidebook* for explanations of the error messages. If that fails, go to education.ti.com and use the TI-Cares™ KnowledgeBase search found in FAQs.

I'm getting a result but it is wrong

Check for

◆ Typing direct letters in place of pasting a command. For example, **SIN(X)** can be entered using the **ALPHA** keys, but it is not the same as **sin(X)** using the **SIN** key.

◆ Parentheses mismatch. Count and match parentheses carefully.

◆ Subtraction versus negative symbol. For example, if the negation sign is used to enter **X-10**, the result will be −10 times the value of **X** and not **X** minus 10.

◆ Correct default settings. For example, check the **MODE** for the **Radian/Degree** angle setting.

My program won't run

Program errors are difficult to diagnose. Scroll and check your code with **PROGRAM EDIT**. Add temporary displays and pauses to check the progress and isolate the problem.

INDEX

NOTES

NOTES

NOTES

NOTES

NOTES

<u>NOTES</u>

NOTES

NOTES

NOTES

NOTES

NOTES